THE MAKI

Lionel Kochan, ...
children, was b... ...
cated at Haber... ...
Christi College, ..., where he was a
scholar, and at the School of Slavonic and East
European Studies. He took a Ph.D. at the London School of Economics, and is now a Reader
at the School of European Studies at the University of East Anglia. He has also published
Russia and the Weimar Republic, *Acton on History*, *The Struggle for Germany 1914–1945*, and
Russia in Revolution 1890–1918. Lionel Kochan
completed his war service in the Intelligence
Corps: he is very fond of chess and most forms
of vigorous outdoor exercise.

THE MAKING OF
MODERN RUSSIA

LIONEL KOCHAN

PENGUIN BOOKS
IN ASSOCIATION WITH
JONATHAN CAPE

Penguin Books Ltd, Harmondsworth, Middlesex, England
Penguin Books Inc., 7110 Ambassador Road, Baltimore, Maryland 21207, U.S.A.
Penguin Books Australia Ltd, Ringwood, Victoria, Australia

—

First published by Jonathan Cape 1962
Published in Pelican Books 1963
Reprinted 1965, 1967, 1968, 1970

—

Copyright © Lionel Kochan, 1962

—

Made and printed in Great Britain
by Hazell Watson & Viney Ltd
Aylesbury, Bucks
Set in Linotype Times Roman

TO
NICK, ANNA, AND
BENJAMIN

Contents

Maps

Foreword

THE aim of this book is to show how the Kiev state of the tenth, eleventh, and twelfth centuries developed into the Soviet Union of today. In order to make this evolution at all comprehensible I have had to be selective rather than comprehensive. I have, in addition, concentrated on the later phases of the subject. Thus, although I have tried to give some idea of the whole development of Russian history, it is the later centuries that are of most importance in the context of my present aim. It is their treatment which is intended to justify the title. As a guiding principle it has seemed to me of primary importance to deal with the growth of the state, political developments, economy, and social problems. Such themes as foreign affairs, literature, and cultural life are mentioned only in so far as they illuminate or help to determine the development of the central themes. I have also written from the point of view of the dominant Great Russian nationality; problems such as those of Poland, the Ukraine, Siberia, and central Asia are touched on only marginally. This applies also to the Communist International. Despite these limitations, I hope I have been able to sketch a framework into which these neglected themes may conveniently and intelligibly be fitted.

Edinburgh   LIONEL KOCHAN

CHAPTER 1

The Rise and Fall of Kiev Rus

THE formative centuries of Russian history had their setting in a vast exposed plain, stretching from Eastern Europe to central Siberia. To the north and north-east, the White Sea and the Arctic Ocean formed the boundary. To the south lay the Black Sea, the Caspian, and the Caucasus mountains. The Urals are no kind of climatic or even physical barrier. Nowhere does their altitude exceed 6,000 feet. They rise from gradual foothills to a mean altitude of some 1,500 feet. Numerous valleys and passes make transit easy, both eastwards and westwards. Where does Europe end? Where does Asia begin? It is impossible to say. It is this geographical indeterminacy that helps to account for the perennial question: does Russia belong to Europe or to Asia, or does it form some complex world of its own?

The distinctive features of the plain are the uplands that form the watersheds of the area's river system. The Valdai Hills, for example, some 200 miles north-west of Moscow, are nowhere more than a thousand feet above sea-level. Yet they are the source of such major rivers as the Western Dvina, flowing into the Baltic; the Dnieper, flowing into the Black Sea; and the Volga, the greatest of all, which empties into the Caspian. These rivers and their tributaries linked the territory they watered to the countries beyond the seas. Portages connected one system with the next. It was at strategic points on these interlocking routes that the first Slav towns developed – Kiev, Novgorod, Polotsk, Chernigov, Smolensk. In the fourth century B.C., Herodotus already knew the Dnieper as 'the most productive river ... in the whole world, excepting only the Nile. ... It has upon its banks the loveliest and most excellent pasturage for cattle; it contains abundance of the most delicious fish ... the richest harvests spring up along its course and, where the ground is not sown, the heaviest crops of grass. ...'

This describes, of course, the fertile southern reaches of the

Dnieper. Such vegetation was by no means uniform throughout the vast Eurasian plain through which the river slowly meandered. In the far north lay a belt of cold, barren tundra. Then comes a well-watered forest zone of enormous expanse, ill-suited to agriculture but nourishing well-nigh inexhaustible numbers of wild animals to provide food, clothing, and furs. Throughout the south lie the exceptionally fertile earth regions, which begin in the Ukraine and stretch eastwards into southern Siberia. These, in their turn, give way to the arid, semi-desert, sandy steppes to the north and east of the Caspian.

When in the sixth and seventh centuries A.D. the Slavs settled on the Russian plain, they were by no means the first peoples to have established themselves there. There is evidence that for a thousand years there had been continuous settlement and occupation by nomadic and semi-nomadic tribes – Scythians, Sarmatians, Goths, Huns, Avars, Khazars. One after another, warlike hordes from the east and north had made their way across the exposed steppe-lands and along the navigable rivers.

This was the world in which the Slavs made their appearance in about the sixth century. They first began to move eastwards from the Carpathians during the great Migration of Peoples. After the death of Attila in 453, and the collapse of the Hun empire, this process quickened and a threefold division of the Slav migratory stream emerged: to the Elbe, Oder, and lower Vistula went the western Slavs – forerunners of the Czechs, Slovaks, and Poles; to the Balkan peninsula went the southern Slavs – Serbs, Croats, Slovenes, and Bulgars; and to the Dnieper, the upper Volga, and the shores of Lakes Ilmen and Peipus, the eastern Slavs – the Russians.

Those tribes of the eastern Slavs who settled along the waterways of the middle Dnieper and the northern lakes were rapidly drawn into the trading network of the area. Their closest neighbours, and also their conquerors, were the Khazars, a people of Turkic origin who had become converted to Judaism. But it was a mild form of conquest, entailing little more than the payment of tribute.

The Khazar Khaganate had the rough shape of a triangle.

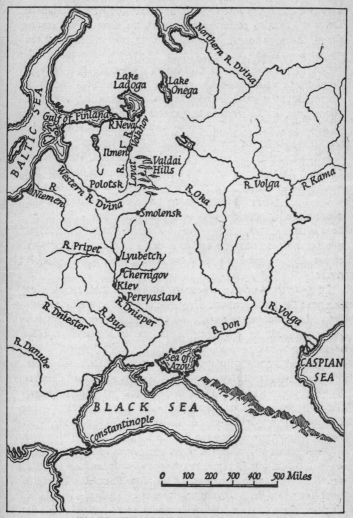

1. WATERWAYS OF EASTERN EUROPE

Its base ran from the Caspian Sea to the Sea of Azov; the two sides rested on the Volga and the lower Don. Originally nomads, the Khazars had later become agriculturists and traders. They were advantageously situated to become a centre for transit trade between the rich Arab Caliphate to the east and commercial Europe to the north-west. From the subjugated Slav tribes came slaves, honey, wax, and furs. From the Trans-Caspian areas, including Bokhara, came silks, textiles, and metals *en route* to Byzantium and Asia Minor. North Africa, Italy, and Spain were also among the recipients of products from Trans-Caspia, routed via the Khaganate.

Into this network of trading relationships the Slavs were drawn. The Dnieper area of Slav settlement was in fact the southern portion of what was later known as the great 'way from the Varangians to the Greeks', that is, from Scandinavia to Byzantium. This route, one of the most important trade-routes of early Europe, began in the far north with the Gulf of Finland; thence to Lake Ladoga by way of the River Neva; by way of the River Volkhov to Lake Ilmen; by way of the Lovat to the vicinity of Veliki Luki. Here the trader took to *terra firma* to reach the upper waters of the Dvina. Above Smolensk, the Dnieper itself was reached. To the north of Zaporozhę there was one further trans-shipment to avoid the cataracts. The west coast of the Black Sea finally led to the great trading and metropolitan centre of Byzantium.

Under Khazar auspices, there also developed independent Slav trading-routes to the east; and both the Khazar capital, Ityl (on the Volga, near the present-day Astrakhan), and Sarkel (on the Don), another important Khazar metropolis, had large Slav populations. One of the chief characteristics of the early Slavs was the large number of towns to which their trading activities gave rise. The number has been variously estimated at between 300 and 600. To the Varangian traders of Scandinavia and the Baltic, the Slav territories were known as *gaardaríki* – kingdom of the towns.

This extreme decentralization was countered in the ninth century by a process of consolidation as each town attempted to seize the territory contiguous to it and dependent on it. The result was incessant strife. The victims were sold into slavery in Byzantium. A special square in Constantinople was set aside for the sale of slaves from the north. It was in these circumstances that the ancient chronicle records the invitation to the Varangians to restore order in the Slav lands: 'Our land is great and rich but there is no order in it. Come, then, to rule as princes over us.' It is in terms similar to these that the *Anglo-Saxon Chronicle* accounts for the arrival of the Angles and Saxons.

In 862, Great Novgorod, on the River Volkhov, one of the most important Slav towns, fell to Rurik, the Varangian chieftain. Rurik died in 879. His successor, Oleg, led a campaign southwards down the river route to Byzantium. He took Smolensk and Lyubetsch, and in 882 Kiev, which he fortified and made his capital.

The Varangians or Norsemen were known to half Europe in the easily interchangeable roles of trader and pirate. They led attacks on the coasts of France, England, and Ireland; they sailed up the Seine, and through the Straits of Gibraltar into the Mediterranean; about the year 1000 one of their forays even reached the coast of North America.

In Rus, as the Slav lands were known to the men of the north, the Varangians, even before the 'invitation' to Rurik, had become familiar with the route to Byzantium through acting as armed bodyguards to the Slav traders. Their assumption of power over the Slav tribes may well be seen as an extension of their previous role. They did not come to build a state; rather they became the ruling stratum of princes over the mass of Slav society. Even this may be an exaggeration. Rus, with Kiev as its later capital, may also be seen as the conquest of the less developed Slav tribes of the south by the more developed ones of the north, aided by Varangian merchant-adventurers and princes. In any event, it was less than a century before the first signs appeared of the absorption of the Varangians by the numerically far superior Slavs. Oleg's successor, Igor, for example, had a son, born in 942, who was given a Slav name – Svyatoslav.

The relationship between the Varangians and the Slavs was

limited in the main to plunder and the imposition of tribute. The Byzantine Emperor, Constantine Porphyrogenitus, describes the Varangian princes and their retinue moving among the Slav villages and homesteads at the onset of winter. They collected tribute in the form of marten, squirrel, and fox furs, honey, wax, and other forest products. The seizure of slaves for sale in Byzantium was another purpose of these winter forays. In summer the proceeds were disposed of, after the journey southwards to Constantinople.

Given this rule at home, the foreign policy of the rulers of Kiev was to keep their trading-routes open. As early as 907, Oleg led a fleet of 2,000 ships, each with forty men aboard, against Byzantium, and was successful in forcing on the Greeks a trade treaty favourable to Kiev. This assured the Slav and Varangian merchants of access to the Byzantine market, accommodation and food at the expense of the Byzantine State, as well as a supply of provisions, sails, and anchors for the return journey northwards. In return, the Slavs submitted themselves to close supervision and control at the hands of their suspicious hosts and fellow traders. They had to be registered, they undertook to refrain from looting, not to appear in groups of more than fifty at the market, and not to carry arms.

Igor's successor, Svyatoslav, developed widespread schemes of conquest. He was evidently a true Viking. He left internal matters to his mother, Olga, reserving to himself the task of expanding to the maximum the frontiers of Kiev Rus. Whereas his predecessors had made Byzantium their main objective, Svyatoslav turned his efforts towards the Volga basin, the route linking Russia with the East. He overthrew the Khazar Khaganate when he took its principal towns, Sarkel and Ityl. He seized the Kuban area in the northern Caucasus and penetrated in force to the Sea of Azov. He came finally to grief after his overthrow of the Bulgarian kingdom on the Danube. When he turned back to Kiev in 973 he and his troops were attacked at the Dnieper cataracts by Pechenegs tribesmen, the most recent nomadic invaders of the steppes. The Slavs were overwhelmed. Svyatoslav's skull, it is said, was made into a drinking-cup for a Pecheneg prince. Never again did the history of Kiev

Rus know a realm that stretched from the Volga to the Danube and northwards to the Finnish Gulf and Lake Ladoga.

The next half-century or so saw the efflorescence of its culture and the emergence of Kiev Rus as an acknowledged European Power. Under the rule of Vladimir, this was considerably facilitated by the adoption of Christianity, in its Byzantine form, in 988. But the preliminaries to the conversion of Rus clearly show the widespread influences to which the State was exposed. To Vladimir came Jews from the Khazar Khaganate, Muslims from a Bulgar state on the Volga, Catholics from Germany, and a Greek philosopher from Byzantium. The latter's victory was no doubt dependent in part on Kiev's commercial relationship to Constantinople.

With Christianity there came an organized Church under a Metropolitan, nominated from Constantinople and usually a Greek. Simultaneously there came into existence manuals of Church law and ritual, volumes of sermons, the lives of the first saints and monks of Kiev Rus, and the first national chronicles. The new religion made its visual impact in the construction of churches in Byzantine style and in the form of frescoes, icons, and mosaics. The use of the icon followed the defeat of Byzantine iconoclasm in the ninth century.

The schism of 1054 between the Eastern and Western forms of Christianity, occasioned by the doctrine of papal supremacy, had as yet no part in separating the Kiev state from Western or Central Europe. Nothing shows more clearly the European acceptance and standing of Kiev than the marriage alliances concluded by Yaroslav, the ruler of Kiev in the second quarter of the eleventh century, and his family. He himself was married to a daughter of the King of Sweden. His sons had German and Polish wives. Three of his daughters, Anna, Elizabeth, and Anastasya, were married to the Kings of France, Norway, and Hungary. Anna, consort of Henry I of France, was able to sign her name – whereas her French husband was illiterate.

On the other hand, the ethos of Byzantine Russia would later differ profoundly from that of Western Christianity. Through the use of the Slavonic vernacular as a medium of communication, it was less able to share in the European culture of Greek and Latin. Secondly, it set greater store on piety than on knowledge, so that the Eastern Church had a pronounced anti-intellectualist colouring. Thirdly, it stood in much more intimate contact with the State than did the Church in almost any country of Western Europe. When the Mongols came to make their own contribution to the development of the Muscovite state, the consequence would be a reinforcement of the collective at the expense of the individual.

The basis of the Kievan economy was agriculture. This applied with particular force to the area around Kiev and Chernigov. Here the climate and soil were both favourable. To the north the abundance of swamplands and wooded areas made agriculture far less productive and predictable. In the south, however, there was ploughing by horse, and the sowing of wheat, millet, barley, corn, rye, flax, peas, and poppies. The early records also make mention of almost every species of domestic animal – horse, bull, cow, dog, goat, lamb, sheep, oxen, and hog.

The hunting and trapping of fur-bearing animals was the speciality of the north. Here the expanse of forest and water was the home of sable, beaver, marten, polar-fox, and fox. Squirrel skins were dealt in by the tens of thousand and were also used as small change. Trade in honey and wax and the fishing industry were other occupations of the north.

At the top of the social hierarchy stood the prince, whose functions were chiefly military and judicial. He had to defend his principality and to supervise the administration of justice. In both capacities he was advised and aided by a council of boyars, i.e. independent landowners, a nascent landed-aristocracy. Urban government was controlled by the *vyeche*, a popular assembly of all free adult males. The towns, and they probably numbered more than 300, were the home of the emergent class of artisans – mainly carpenters, masons, bridge-builders, joiners, saddlers, and the like. These were the men who built the wooden homesteads, market-places, cattle-sheds, and bath-houses that urban life required. There were also metal-workers and artisans, who specialized in the processing of leather and the manufacture of linen and other types of coarse textiles.

Kiev itself was, of course, the paramount town in Rus. Yaroslav in the eleventh century hoped to make of it a Slav Constantinople, and it did undoubtedly become one of the showpieces of the Middle Ages. One visitor counted no fewer than 400 churches, the work of Byzantine architects, stonemasons, mosaic workers, and painters. The town was surrounded by protective walls. Inside there were eight great market-places attended by merchants from all the regions with which Kiev Rus traded – Germany, Bohemia, Hungary, Poland, Scandinavia, Byzantium, and the countries of the Orient. Adam of Bremen, the German chronicler, found Kiev 'the fairest jewel' in all the Greek world.

The bulk of the rural population was composed of *smerdi*. These, during the ascendancy of Kiev Rus, were more or less free farmers or rural householders with their own homestead, animals, ploughland, and primitive farming implements. They were their own masters to the extent of being able to bequeath property, almost without qualification. The Russkaya Pravda – the eleventh-century code of laws which is the chief source of information on the social relations of this early period – describes other rural classes in a complex network of obligation. The *zakupi* were landless peasants whose existence depended on the labour they performed for others, including the *smerdi*. They had no agricultural tools of their own, and used their masters' ploughs and harrows. The *zakupi* did, however, possess independent households of a kind, which they maintained apart from those of their masters. They were legally free, and their dependence theoretically ceased as soon as they had discharged their debt by labour. If they failed in this, or if they took to flight to evade their obligations, they fell into the lowest status of all – that of *kholopi*, who were barely distinguishable from slaves.

Yaroslav died in 1054; and with his death the decay of Kiev set in at an increasing rate. Weaknesses in this first Russian state became all too obvious. Thus, its unity and integrity were endangered by the system of rule that Yaroslav enjoined on his death-bed. In principle, all his heirs – and he had five sons and a grandson – undertook to rule Russia collectively. The eldest heir ascended the throne of Kiev, receiving the title of Grand Prince. His brothers assumed rule over lesser towns and adjacent territories, depending on their position in the family hierarchy; the youngest brother, for example, received the smallest town. But this apportionment was in no wise final. When the grand prince of Kiev died, the throne was assumed by his next surviving brother, and this in turn led to a general 'moving up' among all the members of the ruling dynasty. This system inevitably became a source of dissension among the princes and their families. In 1097 a conference of princes at Lyubetsch attempted to give some order to the system of succession, but no lasting result was achieved. By about 1100 there were some twelve separate principalities on the territory of Kiev Rus, all virtually unconnected by any central power.

These internecine feuds naturally left the country less fitted than ever for self-defence against the latest wave of barbarian nomads from the East – the Cumans. Their raids and depredations, not only on the trading routes but also against villages and homesteads, led to the increasing depopulation of the Dnieper basin. From the Smolensk region, for example, in 1160, Cuman raiders carried off into slavery some 10,000 Russians.

The Crusades were a second external factor tending to undermine Kiev's economic existence. The development of the Mediterranean as a trading highway to the Orient inevitably diminished the importance of Kiev and the Black Sea route to Byzantium and beyond. What Venice and Genoa won, Kiev lost. The capture of Constantinople during the Fourth Crusade in 1203–4, expressly at Venetian instigation, was a powerful blow at Kiev's foreign trade.

In the meantime, other enemies were appearing in the north-west – the Teutonic Knights, the Lithuanians, and Swedes. From the south-west came the Hungarians, pressing eastwards.

To the *smerd* threatened with enserfment, to the free population exposed to wholesale kidnapping, to the traders with their livelihood menaced, the response to deteriorating conditions was flight. One route led south-westwards to Galicia, but in the main the route led north-eastwards, from the Dnieper basin to the region of the upper Volga basin, and the Oka, where new centres of Russian life began to form, based on the natural economy of an environment hitherto peopled by primitive Slavic and Finnish tribes.

The transfer of power to the north-east was dramatically illuminated in 1169. In that year a coalition of twelve princes under Andrei Bogolyubsky sacked Kiev; and Andrei removed the capital of Rus to the township of Vladimir on the Klyazma River, in his native principality of Rostov-Suzdal.

But Vladimir was no more the capital of a united country than Kiev had been. When the Mongols invaded Rus in the last thrust westwards of the Eurasian nomads – successors to the Scythians, Sarmatians, Huns, and Cumans – they found a dismembered country before them. Their conquest of 1237–40 decisively broke Slav contact with Europe, or at least Western Europe, and turned the country eastwards. Not until the six-teenth and seventeenth centuries, for example, did such former centres as Kiev and Smolensk rejoin the Russian State. The Dnieper basin was replaced by the more remote region of the Oka and the Volga as the dominant locale of Russian history. It was in this area under the aegis of the Mongols that the small principality of Moscow asserted its position as nucleus of the future Russia.

CHAPTER 2

The Mongol Conquest and the Rise of Muscovy

IT was in 1223 that the Mongols first swept into Rus. By 1240 they had conquered in two major invasions all the Slav principalities, and in 1242 they established their headquarters at Sarai, on the lower Volga, not far from where Volgograd stands today. Thereafter, for the best part of two and a half centuries, they dominated political and economic life over a vast area of Eastern Europe. By the end of that period two new forces had come into existence – a Lithuanian state, stretching from the Baltic to the Black Sea, and Muscovy, an inland principality, centred on Moscow, and dominant over all the other Slav principalities.

This was a situation very different from that which the Mongols had encountered at the beginning of their sway. Then they had found perhaps a dozen Slav states, from Novgorod in the north to Ryazan in the south, with no tradition of unity that would enable them to combine against the invader. What was the nature of Mongol dominance on this conglomeration of states? The Khans ruled indirectly and made no attempt to interfere with the internal affairs of the Slavs except where and when their supremacy was threatened. They exercised their power, not by colonization or by imposing their own system of government, but by levying tribute on the Slavs and by reserving to themselves the right to confirm in office each new ruler in each of the principalities. Before any ruler could ascend his throne he had to travel to Sarai, and there receive, or more likely intrigue for, his *yarlyk*, or authorization. More than 130 princes made this pilgrimage. Many took their families with them. It was customary for the prince to draw up his will when he left for Sarai, lest he fail to return alive. Economically, Mongol overlordship expressed itself in the levying of taxes and conscription. Mongol officials and census-takers were soon at work throughout the whole of Rus, listing all adult males on

SWEDES

BALTIC SEA

LETTS

TEUTONIC KNIGHTS

LITHUANIANS

R. Néva

Novgorod

K I E V   R U S

Suzdal

Vladimir

R. Volga

Kiev

Galitch

R. Dnieper

POLOVTSI (after 1055)

PECHENES (until 1036)

KHAZARS (until c. 1000)

MONGOLS (13th century)

HUNGARIANS

R. Danube

BULGARS

BLACK SEA

Constantinople

MONGOLS (13th century)

CASPIAN SEA

BYZANTINE EMPIRE

0   100 200 300 400 500 Miles

2. KIEV RUS AT ITS HEIGHT, AND ITS CHIEF
ENEMIES

their paper scrolls. They imposed a fixed *per capita* tax, known as *yassak*, and also tolls and duties on salt, ploughs, bridges, ferries, and the movement of cattle across internal boundaries. Mongol troops quartered in the chief towns of each principality bloodily suppressed the frequent Slav uprisings. Modern Russian, in its use of words of Tartar origin to express such ideas as 'goods', 'money', 'toll', 'exchequer', shows the impact made by Tartar financial depredations during these early centuries of the people's history.

In fact, for all the indirectness of the Mongols' rule, they did exert considerable direct influence – and for the most part in a destructive sense. During the period of their dominance there was decline and degeneration in the level of craftsmanship, in artisanry, in the position of the town, which lost its free urban institutions, in the artistic level of the chronicles and religious writings, and politically, in the increasing fragmentation of the political unit. Above and beyond all this, the rule of the Mongols cut Rus off from the rest of Europe. Only Novgorod and Pskov still enjoyed contacts with the north German towns. As Mongol rule weakened, some contact did again develop with the Balkans. But when Constantinople fell to the Turks in 1453 and there was a spread of Turkish influence, this access to the West was blocked. There is no doubt that Russian cultural backwardness in the centuries following the collapse of Mongol rule was intimately related to the impact of that rule. From one point of view, the country's subsequent history can in fact be seen as a prolonged attempt to regain contact with the West.

What Mongol rule did contribute to Russian life was a certain absolutist and autocratic framework. By far the most important effect of the Mongols was to further the creation of the Muscovite type of autocracy that emerged at the end of the fifteenth century. What the Russians at first dreaded they later imitated, as a measure of self-defence. 'Autocracy and serfdom were the price the Russian people had to pay for national survival', a recent historian has noted.[1]

But this is to anticipate. It took some two centuries – from the mid thirteenth to the mid fifteenth – before Muscovy had won

1. George Vernadsky, *The Mongols and Russia*, Oxford, 1953, p. 390.

for itself undisputed supremacy over the other Slav states of north-eastern Russia as the paramount successor-state to the Mongols. It has been said:

The history of Moscow is the story of how an insignificant *ostrog* [blockhouse] became the capital of a Eurasian empire. This insignificant *ostrog*, built in the first half of the twelfth century, on an insignificant river by an insignificant princeling, became, in the course of time, the pivot of an empire extending into two, and even three continents.[1]

The chronicles first mentioned Moscow in 1147. It was then nothing more than a village domain belonging to the prince of Rostov-Suzdal, and evidently a defence outpost against the south. In 1237 it was sacked by the Mongols and razed to the ground. In 1263 it re-enters history as the permanent capital of a minor principality ruled by Daniel, the youngest son of Alexander Nevsky, conqueror of the Swedes and Livonian Knights. ('Nevsky' is derived from 'Neva', the name of the river on which Leningrad stands, the scene of Alexander's victory over the Swedes.)

Moscow was at this time far inferior in importance to any of the other principalities such as Ryazan, Tver, Smolensk, Vladimir, and Pereyaslavl-Zaleski. But certain objective factors would eventually make it the nucleus of a national Russian state. There was the geographical factor, for example, as in the case of Kiev. Moscow, situated on the river of the same name, enjoyed a sheltered position that was of inestimable advantage in the thirteenth and fourteenth centuries. At a time when the west and south-west were harassed by an expanding Lithuania which eventually reached the Black Sea, when the trans-Volgan territories were the prey of freebooters from Novgorod, barely distinguishable from pirates, and when the other outlying territories of Rus to the south and east were the repeated victims of Tartar raids, the site of Moscow offered shelter and protection. No serious attack was made on it in the century and a quarter that separated a Tartar raid of 1238 from a Lithuanian

1. R. J. Kerner, *The Urge to the Sea*, Berkeley, California, 1942, p. 35.

foray of 1368. Moscow became a centre of settlement for the uprooted subjects of the more exposed principalities.

Moscow also drew strength from its position in the Russian Mesopotamia, in the *Mezhduryechie*, the area bounded by the upper Volga and the Oka. The axis of the future Muscovy was to be the Baltic-Volga-Caspian trade-route. But before this stage was reached, Moscow already enjoyed a geo-commercial location on which, fundamentally, its expansion depended. Lying as it did on the Moscow River, flowing from north-west to south-east, and linking the system of the middle Oka with the system of the upper Volga, the principality, and later grand-duchy, was thus connected with all the important river systems of northern and western Russia. Through a complex of portages and tributaries Moscow had access to the Volga on the north, east, and south, and to the Dnieper and Western Dvina on the west. Farther afield, waterways led to the Baltic and the Black Sea. Moscow was thus situated at the centre of the two supremely important trade-routes and waterways crossing Rus from north to south and from west to east – the Baltic-Caspian and the Western Dvina-Volga.

But Moscow's central position at the crossroads of trade demanded, for its full exploitation, that it extend its direct control over the whole *Mezhduryechie*. It would otherwise have been dependent on the goodwill or acquiescence of a multiplicity of territories, large and small, for its continued existence. Moscow must either dominate or suffocate.

In the event, the forces making for domination were by far the stronger. The strategy of all the early princes and grand dukes of Muscovy was aimed at winning control of the portages and rivers that would eventually give access to the open sea. A dominant theme of Russian history was already evident in its earliest years.

A variety of methods was practised, few of them pretty. There was purchase, outright conquest, acquisition by diplomatic means, and the conclusion of treaties with petty princelings that left them masters in their domain but bound to Moscow by ties of service. Finally, there was colonization. The various processes were all in motion at the very beginning of the fourteenth century. When Daniel, the first ruler of Moscow, died in 1303 he was able to bequeath to his successor a principality that had doubled its area in a single reign. From that point, by and large, the history of Moscow shows a record of steady expansion. Among the earliest acquisitions were Mozhaisk and Kolomna. The first lay at the source of the Moscow River, the second at its junction with the Oka. The two towns thus gave control of the whole course of the river. Success here opened the way to further penetration northwards. Ivan Kalita (='money bags'), who ruled Moscow from 1325 to 1341, bought control of the important trans-Volgan territories of Uglitch, Beloozero, and Galitch. These were not incorporated in his domains, but their rulers thereafter retained their positions by virtue of the grace of Muscovy.

Ivan strengthened this dependence by his policy of ransoming Slav prisoners from the Mongols and subsidizing their settlement on his newly dominated territory. In the second half of the fourteenth century and the first half of the fifteenth, Muscovy completed its most impressive phase of expansion. The seizure of Vladimir and Starodub gave command of the Klyazma River, and the gradual erosion of the independence of Nizhni Novgorod and of other eastern principalities set Muscovy athwart the Volga itself. Nizhni Novgorod, the gateway to the lower Volga, brought Moscow within striking distance of the road to Asia, with its particular prize the silk trade of the East.

The slow emergence of Muscovy as the paramount Slav principality was bound to influence the Slav-Mongol relationship. Ironically, Mongol policy actively encouraged the growth of the very power that would eventually bring about the eclipse of the Mongols. No Slav prince was more assiduous in courting the reigning Khan, or more humble in his diplomacy, than the prince of Muscovy. Cash, at Sarai, was all but omnipotent. This weakness the wealthy rulers of Muscovy were well fitted to exploit. They also derived much of their influence at Sarai from the share they took in crushing any movement hostile to the Mongols among the lesser Slav principalities. The troops of Muscovy functioned more than once in this role. It was also

not infrequent for Muscovy – no less of course than its rivals –
to use Mongol troops in this struggle against the other principalities.

It was through such devious methods that, as early as the
reign of Ivan Kalita, Muscovy acquired the undignified but
influential rights of collecting the taxes imposed by the Mongols on the Slavs. This gave further pressure to Muscovy's
financial squeeze. Moreover, in 1353 the principality became
acknowledged by the Tartars as the judicial authority over the
other princes of Rus.

This gradual accretion of strength and influence resulted, in
1380, of the first serious Slav attempt to cast off the Mongol
rule. At the field of Kulikovo near the Don the whole of
northern Rus, led by Dmitri Donskoi of Muscovy, fought and
won a pitched battle against the overlord. But, though the
victory was striking, it was in no way conclusive. Barely two
years later the Mongols returned in force and utterly devastated
Moscow. Vladimir, Mozhaisk, and other towns were similarly
laid waste. In 1408 a renewed Mongol invasion led to further
widespread ravages in Nizhni Novgorod, Rostov, and Serpukhov. But despite these, and despite losses to Lithuania in the
west and an outbreak of internecine feuds in Moscow itself,
the dominance of the grand duchy was firmly established by
the middle of the fifteenth century.

CHAPTER 3

The Formation of a National State

IVAN III, also known as Ivan the Great, ruled Muscovy from 1462 to 1505. For the first time a clear personality emerges from the ruck of warring princes and grand dukes. Contarini, a Venetian traveller, describes Ivan in a report of 1476 as tall, lean, and handsome. Some forty years later another visitor to Moscow, Baron Herberstein, an envoy of the Holy Roman Emperor, was told that Ivan's mere appearance was enough to send into a swoon any woman who encountered him unprepared. If, writes Herberstein, drunkenness caused him to fall asleep during a banquet – and this was not infrequent – his guests would await in fear and trembling his return to consciousness, when he would resume his jocularity.

In political and military matters Ivan avoided frontal combats. He preferred to achieve his aims through calculated and devious diplomacy. Force and violence he used as a last resort.

As the creator of a centralized state, Ivan has been well compared with his contemporary, Louis XI of France. His reign was without doubt a watershed in the development of the future Russian State. For all the conquests of his forerunners, he found Muscovy flanked to the north by the great commercial republic of Novgorod, which dominated a vast area from the Gulf of Finland to the Urals. To the west stretched the combined Polish-Lithuanian empire, lying along the Western Dvina, the Dnieper, and the Black Sea. South of Tula and Ryazan lay a vast expanse of steppe-land, as far as the Black Sea and the Caspian, dominated by the Tartars of the Crimea and the lower Volga. To the east, the Tartar Khanate stood athwart the Volga.

Apart from the absorption of lesser Slav principalities such as Yaroslavl, Tver, Rostov, and Ryazan, it was no longer possible for Muscovy to advance farther without meeting head-on opposition from powerful empires. It is a measure of Ivan's success that by the end of his reign the Mongols were no longer a threat. They themselves, on the contrary, were now on the defensive. Novgorod and the other Slav states had lost their independence, and battle had been joined with the Lithuanians and Poles for expansion to the west.

Internally also, the growth of the autocracy with a Pan-Russian ideology, and the emergence of a class of military serving landowners, mark the second half of the fifteenth century as a turning-point in Muscovy's political and economic development. In Toynbee's words: 'The offshoot of Orthodox Christendom in Russia entered into a universal state towards the end of the fifteenth century, after the political unification of Muscovy and Novgorod.'

When Ivan the Great came to power three main tasks stretched before him: to complete the 'ingathering' of the Slav lands, including Novgorod, under the standard of Muscovy; to win back from Lithuania the south-western lands that had once been part of Kiev Rus; and to complete the overthrow of the Mongols.

The first of these took up the early years of the reign. Ivan strode ahead along the same path that had been marked out since the early years of the fourteenth century. In 1463 the princes of Yaroslavl submitted to Muscovy, lost their independence, and accepted service with Ivan. In 1472 it was the turn of Perm, with its lands on both sides of the Kama River. Two years later the Rostov princes sold what remained of their territory to Moscow. Tver, at one time the most hostile to Moscow of all the Slav groupings, was undermined from within by the defection of its boyars to Moscow. An alliance between Tver and Lithuania brought the conflict to a head, and in 1485 the town of Tver fell without a blow to Ivan's encircling army. The ruler fled to Lithuania and Galicia, never to reappear on Muscovite soil. This same process was continued by Ivan's successor, with the absorption of Pskov and Ryazan.

Of all these accretions of power the most important was Novgorod. Had Novgorod not been crushed, how could Muscovy ever have become consolidated into an independent Pan-Russian State? Muscovy was landlocked: Novgorod held the key to the Baltic and the open sea. This seaward surge was to be one of the most dominant themes of Muscovite, and later Russian, foreign policy for the next two and a half centuries. There is a steady consistent theme that links Ivan III in the fifteenth century with Peter the Great in the eighteenth.

The struggle for Novgorod reached its culmination in Ivan III's reign. It had long been in the making. Repeated efforts to dominate Novgorod had been made over the past century and a half by the Muscovite princes. As far back as the days of Ivan Kalita, Novgorod had been forced to pay tribute to her southeastern neighbour and to accept foreign governors (who did not, however, intervene in the republic's internal affairs). But whereas the stakes had originally been river routes and furs, they were raised, when Lithuania emerged to press Novgorod on the west, to nothing less than the continued existence of Muscovy.

The fact is that the Novgorod policy of manoeuvring between her two powerful neighbours could succeed only so long as the latter were in relative equilibrium. Not only was this threatened by the Muscovite impetus; the position was further complicated by the religious struggle between an aggressively Catholic Poland and Lithuania, and an equally aggressive Orthodox Muscovy. Several times Ivan warned the Archbishop of Novgorod to remain loyal to the Orthodox Church and to reject all contact with Catholicizing tendencies emanating from Kiev, seat of the Catholic Metropolitan Gregory. (The latter had been consecrated by the Pope, Calixtus III.)

There was also an open class struggle inside Novgorod itself. The boyars and oligarchs inclined to Lithuania, an aristocratic state; the artisans and lower classes to Moscow, which already had some reputation as a national state, where the right of private property, as personified by the boyars, had to yield to the 'national' need.[1]

Early in 1471 the pro-Lithuanian party inside the city came to an agreement with Casimir IV, King of Poland, and grand prince of Lithuania. Casimir promised to respect the Orthodox Church and to support Novgorod against Moscow. The treaty

1. See p. 37, where this point is discussed more fully.

also entitled Casimir to install a Polish governor in Novgorod on condition that no change was made in the constitution.

In May 1471, Ivan and his Council of War decided to take up the challenge. Novgorod's new alliance enabled him to proclaim a religious war, a sort of crusade. The campaign took in not only Novgorod territory proper but also the far north. By August 1471 all resistance was over. An unexpectedly hot summer, which dried up the intervening swamps, helped to accelerate victory – as did also the half-hearted resistance put up by the Novgorod plebs. Many saw in the autocrat of Muscovy, at odds with his own boyars, a liberator from their overlords. In the battle along the River Shelonna, for example, a Muscovite force of some 4,000 or 5,000 men overcame a Novgorod force many times stronger in number.

The peace that Ivan imposed was comparatively mild. He cancelled the treaty with Lithuania; an indemnity of 15,000 roubles was exacted; he forced Novgorod to yield up to Muscovy certain of its colonies and also to recognize his, Ivan's, sovereignty. This was not quite the end of Novgorod's independence, but there remained little left to lose. The end was a protracted affair and came after almost two decades of pro-Lithuanian riots and tumults inside the city and half-hearted attempts to reassert a claim to independence. Ivan fomented some of the disorder from outside in order to provide a pretext for intervention. He was also able to interfere with Novgorod's grain supply. By 1489 all was over, with the deportation of many thousands of Novgorod's boyars, merchants, and landowners. Their confiscated estates were bestowed on lesser boyars and other people from Muscovy who undertook, in return for the land, to render military service. As an independent political unit, Novgorod the Great no longer existed. It disappeared as a factor in European trade in 1494, when Ivan arrested all the Hanseatic merchants on Novgorod territory, closed their yard and church, and confiscated all their goods. Muscovy now enjoyed unchallanged sway over a vast fur-bearing empire that stretched from Karelia, on the Gulf of Finland, to the Urals.

The subjugation of Novgorod left Ivan free to deal with his

two other main opponents – Lithuania and the Mongols. This was a complex, interdependent problem, with Ivan manoeuvring uneasily between the west and the south, endeavouring to prevent a union of the two fronts. His technique of exploiting his enemies' divisions could be used to better advantage than ever, now that the centralized Mongol empire had begun to disintegrate. At least three separate centres had replaced the former empire. There were Mongol Khanates at Kazan and in the Crimea, and a third group located between the Don and the Dnieper, the heir to the main Golden Horde. The earlier position was thus reversed. The united Mongols had once confronted the disunited Slavs; it was now a dismembered Mongol empire that came to grips with a united Muscovy.

In the 1460s Ivan had begun to withhold his due tribute from the Mongols. Strengthened by an alliance with the Crimean Tartars against the Golden Horde, he acted as though no dependent relationship existed. Ahmed, the Khan of the Golden Horde, twice summoned Ivan to his capital – without response. The climax came in 1480 when, by agreement with Lithuania, Ahmed launched a campaign against Muscovy.

Two and a half centuries of Mongol domination ended in a strange anticlimax. The two forces came to face each other across the banks of the Ugra River, the south-west frontier of Muscovy. There was one early skirmish – and nothing more. The help that Ahmed was expecting from Lithuania never materialized. This made any full-scale campaign against Muscovy impossible. He had no choice but to withdraw his troops. But because Ivan for his part feared the possibility of a Lithuanian attack, he dared not involve himself in a war with the Mongols, not even when they were retreating. Ivan also withdrew from the banks of the Ugra. All along, he had in fact played a very cautious, evasive role; so cautious as to earn the reproach both of the clergy and of the populace. At one time he had even prepared to evacuate his family and treasure from Moscow to the north. The fear inspired by the Mongols still bit deep.

The tragicomic end to the last attempt of the Mongols to restore their suzerainty was underlined in 1502 when the Golden

Horde itself ceased to exist as a cohesive force. It dwindled into the Khanate of Astrakhan. The Mongols still remained a local threat, of course – from the Crimea an invasion was mounted in 1521, and again half a century later – but never more would they threaten the actual existence of Muscovy, let alone assert a claim to overlordship.

Ivan had now dealt with both Novgorod and the Mongols. There remained only Lithuania and the recovery of the lands of Kiev Rus. Here Ivan was less successful. A number of campaigns in the 1490s and early 1500s pushed Muscovy's frontier westwards to the upper stretches of the Western Dvina, and the Dnieper. This brought Smolensk and Kiev within range. But the Baltic was as elusive as ever.

THE CHARACTER OF MUSCOVY

What sort of Muscovite state was it that emerged from the long struggle with the Mongols and the lesser Slav principalities? It can be seen in so many ways – as the victory of the forest over the steppe, as the transition from a society of independent principalities to an absolutist autocracy, or, economically, as an evolution from a system of diffused landholding and power to a concentration of land and power in the hands of an autocrat backed by a dependent class of service-landholders. But one fact is clear: the ultimate defeat of the Mongols by no means put an end to Mongol influence. On the contrary, the departing Mongols left an indelible impress on Muscovite life and society.

This may be seen in such matters as Muscovite military organization, Muscovy's method of collecting taxes, its criminal law, its diplomatic protocol. Not for nothing had Muscovy fought the Mongols or collected their taxes or applied their legal sanctions for the best part of two centuries. But more important than all this was the impetus that Mongol rule gave to the development of the Muscovite autocracy. This was the kernel of Mongol influence. The Mongol empire was not only erected on the principle of unquestioning, unqualified service to the State; the Khan also demanded the complete obedience of

BARENTS SEA

SWEDEN

BALTIC SEA

G. of Finland
ESTONIA
LIVONIA
COURLAND

Novgorod

Archangel

Northern R. Dvina

R. Pechora

MOUNTAINS

·Vilna

POLAND

Warsaw

LITHUANIA

Kiev

R. Dnieper

R. Danube

Sea of Azov

BLACK SEA

R. Volga

Moscow

Smolensk

Kazan

R. Kama

URAL

R. Ural

R. Don

R. Volga

Astrakhan

CASPIAN SEA

■ Muscovite Principality
at end of 13ᵗʰ century

▨ Territory of Vassili II (d.1462)

⦂⦂ Territory conquered by Ivan III (d.1505)

▨ Further advances in the
reign of Ivan the Terrible (d.1584)

0 100 200 300 Miles

3. THE GROWTH OF MUSCOVY, 1300–1584

the individual to the State. The Khan alone embodied the State, and all those who shared in it (through, for example, the 'possession' of land or the holding of office) did so on sufferance and not as a matter of right. Not all these consequences, of course, had as yet made themselves felt in the reign of Ivan III. But already they were all evident in a more or less developed state: '. . . the increasing regimentation of Muscovite society from the time of Ivan III onwards, the imposition of compulsory service to the State, and the assumption by the Muscovite sovereigns of the political inheritance of the Golden Horde owed much to Russia's resistance to, and counter-offensive against, her Mongol overlords.' [1] It was this specific Mongol colouring that distinguished Muscovite absolutism from that in the contemporary West – in England, say, or Spain.

Ivan III, for example, was the first ruler of Muscovy to claim ownership of all the Russian lands; he vastly extended the internal authority of the grand prince; and he asserted the position of the autocrat as military leader.

The Church, because of its large landholdings, had played an active part in backing Muscovy's claim to Slav overlordship. From the fourteenth century onwards, the association of Church and State, which was much more intimate than anything comparable in Western Europe, had become closer, until by the early sixteenth century the Tsar had come to be considered a semi-sacrosanct personality with unlimited power, the earthly representative of God. To quote Joseph of Volokolamsk, the influential early-sixteenth-century Abbot: 'The Tsar is in nature like to all men, but in authority he is like to the highest God.' It was thus, taking all these factors together, that the autocracy became, in theory, the divinely ordained fountainhead of an undifferentiated concentration of authority – political, in that the Tsar was the only political authority; economic, in that he claimed ownership of the totality of the land; military, in that he led the country in war; religious, in that he ruled by divine right and was committed to maintain and defend the rights of Orthodoxy.

This truly momentous conception, even if for the present it

1. Dmitri Obolensky, *Oxford Slavonic Papers*, v, Oxford, 1954, p. 30.

was more a matter of theory than of practice, produced equivalent changes throughout the hierarchy of Muscovite society. The first to suffer were the boyars. In the early days of Muscovy, or any of the other principalities for that matter, the boyars had been the prince's advisers, of no very different status from the prince himself. The prince had been only the first among equals. The boyars were not his vassals in the sense of Western feudalism. They were independent freeholders; and this was signalized in the contractual relationship that existed between prince and boyar which usually contained the condition: 'the boyars who dwell among us shall be at liberty to come and go'. The boyar could have his lands in one principality but serve a prince in another.

It is easy to see how the rise of Muscovy to sole dominance affected not only the boyars but also the princes who were displaced by Muscovite conquest or purchase. They lost all opportunity to move. Where could they go? To Lithuania or Poland, perhaps? But that was tantamount to treason. Such princes and boyars did not indeed lose their rights as landowners, although their rights were somewhat circumscribed by Ivan. What happened was that by losing their right to depart from service they dwindled into the category of 'serving princes'. They could no longer choose whom to serve but had to serve Muscovy.

The status of the boyars was further jeopardized by Ivan's policy of creating a new class of dependent servitors. The centralized government of the reign, combined with the growth of an army to defend and expand the new state's frontiers, created the demand for a class of bureaucrats and soldiers. Their reward took the form of land grants held on conditional tenure for life. These were the so-called *pomestiya*. The *pomeshchiki*, as such landholders were known, owed their position and their land to their dependent position *vis-à-vis* the Tsar. They held their land only so long as they served. This policy was first applied in the 1480s in conquered Novgorod, whence many thousands were deported in order to make room for 'service men' from Moscow. This whole practice may also in part be an indirect inheritance from the Mongols, for they had known it centuries before its appearance in Muscovy.

But for all Muscovy's indebtedness to the Mongols in matters of class structure and power, it was Byzantium that continued to give the new state its cultural superstructure. This was all the more the case on Ivan's marriage in 1472 to Zoë (known in Russia as Sophia) Palaeologus, the daughter of Thomas Palaeologus and niece of the last Byzantine Emperor. In 1453, when the Turks conquered Constantinople, Sophia and her father took refuge in Rome. In papal circles the plan then evolved of a marriage between Sophia and Ivan, in the hope of bringing the latter into communion with Rome in the sense used by the Council of Florence of 1439, which had sought to reunite the Eastern and Western Churches. On the Muscovite side, Ivan's master of the mint, an Italian named Giovanni Battista Volpe – known in Moscow as Ivan Fryazin (= Frank and, by extension, Westerner) – drew Ivan's attention to the Byzantine heiress in Rome.

The Tsar's first wife had been a member of the ruling house of Tver. Her widower quickly grasped the potentiality for enhanced prestige in the proposed alliance. It would raise him to equal status with the ruling houses of Western Europe. Accordingly, in 1472 a Muscovite mission led by Ivan Fryazin travelled to Rome and returned with Sophia, by way of Lübeck, Reval, Pskov, and Novgorod. Among the company was the Papal Legate; his mission was unsuccessful. In fact, it is said that when he was seen to bear the Latin cross at the head of the Princess's cortège, the Metropolitan Philip, Ivan's spiritual adviser, threatened to quit Moscow in protest.

But if the attempt at religious infiltration was without consequence, the same can by no means be said of the implications of a Byzantine princess on the throne of Muscovy. It was during the reign of Ivan and his son, Vassili III, that Moscow came to be referred to by spokesmen as the 'Third Rome', and the Russians as the 'New Israel' of the Church. Philotheos, a monk from a Pskov monastery, wrote to Vassili III that the Church of ancient Rome fell by reason of heresy, the churches of the Second Rome, Constantinople, fell to the grandsons of Hagar; '... two Romes have already fallen but the third remains standing and a fourth there will not be.' The hopes of all the Chris-

tian empires were now centred on Muscovy, concluded Philotheos. In actual fact, the hopes of Ivan and his successors lay
more in the West than in any Byzantine political legacy. But for
all that, the ecclesiastical doctrine gave an ideological hallow-
ing to the new state.

This new status enjoyed by Ivan III clearly called for an enhanced physical setting. A number of Italian artists and craftsmen had already been included in Sophia's entourage in 1472.
But not until 1475 did Ivan seriously concern himself with the
remodelling of Moscow consistent with its new dignity as the
successor to Constantinople. It was then that he summoned
from Italy five architects led by the famous Bolognese, Aristotle
da Fioraventi. The latter was entrusted with the rebuilding of
the Cathedral of the Dormition, originally built in the Kremlin
in the days of Ivan Kalita. Both here and in the city's other
churches, there developed a blend of Byzantine plan, Russian
onion-shaped domes, and minor Italianate architectural features. The Tsar's residential and official quarters were also
complemented by the construction of two stone buildings – the
Palace of Facets, used for Court ceremonies and the like, and
the Terem Palace, for the royal family's private occupation.
The walls of the Kremlin were strengthened and furnished with
towers and gates.

But Ivan the Great's outward panoply, his success in foreign
policy, his establishment of the autocracy, and his consolidation
of a national Russian State – all these were but the prelude to
the upheavals in the reign of his grandson, Ivan the Terrible.
The boyars did not give in without a fight; and the policy of
creating *pomestiya* wholesale, rapidly brought on a land crisis –
the immediate cause of so many of the convulsions affecting
Russian history, in the sixteenth century no less than in the
twentieth.

CHAPTER 4

The Age of Ivan the Terrible

IVAN THE GREAT'S son and grandson, Vassili III and Ivan the Terrible, inherited both his throne and his problems. Muscovy was still a state that had not yet found a stable social form, was still in process of transition towards an autocracy. How to curb the boyars at home and the Lithuanians and Mongols abroad? How to further the new *pomestiya* system and ensure labour for the new estates? How also, and this is true *par excellence* of Ivan the Terrible, to overcome the backwardness inflicted by the Mongol conquest and establish contact with the more advanced West?

For all Ivan's efforts at solving these problems, or perhaps because he undertook too much, the State of Muscovy at the end of his reign had become so weakened that barely twenty years after his death it had sunk to a nadir of misfortune. In the early seventeenth century, Muscovy stood on the verge of dissolution brought on by a combination of foreign intervention and an unresolved class struggle. But this sinister sequel does not mean that Ivan's reign did not mark a supremely important stage in the growth of the autocracy and in the formation of the dominant class of landed serving gentry, the *pomeshchiki*.

The future Ivan IV was three when his father Vassili died in 1533. From his earliest years the boy's environment nourished a hatred for the boyars that would later coalesce most forcefully with the needs of his internal policy. Elena, Ivan's mother, ruled as Regent during her son's minority. The throne rapidly degenerated into a centre of wild violence, intrigue, and denunciation, as the incipient civil war that had developed in Vassili's reign found open expression. The rival boyar families of Shuisky, Belsky, and Glinsky brought their feuds into the Kremlin itself. One night the Metropolitan himself took refuge from the tumult in Ivan's bedroom. The young Ivan was made to see one of the Shuiskys stretching himself out on the bed in which his father had died. Uproar at the centre was matched outside the Kremlin by corruption, plundering, and pillaging on the part of the magnates. One dominant theme of Ivan's reign – the struggle of the autocracy against the boyars – was early established. The events of the Regency glaringly illustrated the boyars' incapacity.

But for all the personal colouring that Ivan's animus and vengefulness gave to this conflict, it was at bottom economic and agricultural. Who would rule in Muscovy? This was the momentous issue – even, perhaps, whether Muscovy would continue to exist. If powerful and independent landowners could nullify the central power, would not the country eventually go the way of Poland? There, as also in Lithuania, landowning magnates with vast compact blocks of estates were in a position to wield independent power – a standing warning to the would-be autocrats of Muscovy.

Ivan the Terrible was crowned, with full Byzantine rites, in 1547, at the age of seventeen. His titles included the designation 'Tsar', a Slavonic form of 'Caesar'. Ivan was the first Muscovite ruler to make official and regular use of the term, as distinct from its haphazard and casual use at the hands of his father and grandfather. In the same year he also married. He chose his bride, Anastasya Zakharin, from a parade in the Kremlin of several hundred virgins, brought together from all over Muscovy. The girl's family was minor Muscovite nobility. She was also connected with the Romanovs. Her marriage marked the future dynasty's first step to the throne. Sadism was already a marked feature of her consort's character. In his teens he would throw hunting-dogs off roofs or have enemies cruelly put to death, without a qualm. But even with such a husband, the marriage was undoubtedly happy. Deep grief overcame Ivan when Anastasya died an early death in 1560.

The final reduction of the Mongol Khanates of Kazan and Astrakhan marked the new Tsar's first decade. Ivan made several sorties against Kazan before success finally came in 1552. Ivan took the field with 150,000 men and 150 cannon. By building a fortress in the immediate vicinity of the city and using the Volga to float their troops downstream, the men of

Muscovy had soon besieged Kazan on all sides. An engineer from Germany blew up the enemy's water supply system and also used explosives to force a breach in the Mongols' fortifications. The capture of the city was a veritable blood-bath that moved its young conqueror to tears. All the Moslem monumental edifices were torn down and replaced by churches and monasteries. In 1556 Astrakhan followed Kazan in subjection to Muscovy. Now the whole course of the Volga, from its source in the north to its mouth on the Caspian Sea, was in Ivan's hands. New trade-routes to the East, to Persia and India, were opened up to Muscovite traders, a not unimportant development at a time when mercantile capital was beginning to be a feature of the country's economic development. Muscovite traders and settlers could also now cross the Urals into Siberia. Ivan's successes, hailed as revenge for the long Mongol yoke, passed into folk-song and were the occasion for grandiose celebrations in Moscow. Henry Lane, one of the first English merchants in Muscovy, noted a year or so later: 'I think no prince in Christendom is more feared of his own than he is, nor yet better beloved.'

These early years were the honeymoon period of Ivan's reign. In the early 1550s he was even able to carry through by legal means a number of clearly defined anti-aristocratic measures. Thus he curbed to some extent the operation of the hierarchical principle in allocating the higher army commands to the boyars. This measure fully accorded with Ivan's low opinion of the boyars' military capacities. To the same category belongs Ivan's reform of provincial administration and government. Hitherto, the Kormlenie system (from 'korm' – fodder, maintenance, salary) had positively encouraged corruption and maladministration, for the Tsar's *kormlenshchiki* (i.e. 'feeders') and governors had a strong and direct financial interest in the operation of their offices. Thus, much of the proceeds of the fines and dues imposed under this system of local government went directly into the *kormlenshchik*'s pocket; only a small part went to the State treasury.

It was this situation that the reforms of the 1550s attempted to rectify. They tended to dispense with the *kormlenshchik* altogether by creating an independent apparatus of jurisdiction in the countryside. His powers were clipped by making his decisions subject to appeal; the pursuit of runaway peasants was removed from his purview; local assessors were elected to attend his courts; and 'land courts', formed in the main of the lesser gentry, collected taxes directly on behalf of the State. These were used to pay the new officials and also to compensate the *kormlenshchik* for loss of office. The boyar now had independent jurisdiction only over the lesser crimes committed by his own peasantry.

Such measures affected only individual boyars, not their position as a class. Hence the reaction of the magnates was relatively limited. The transition to open hostility came, in the first instance, with a decree of 1550 that created a small standing corps of 1,078 members of the lesser gentry. They were provided with estates in the vicinity of Moscow where they had always to be on hand for whatever military, diplomatic, and administrative tasks the Tsar might give them. This deprived the boyars of certain of their valuable prerogatives.

But this was unimportant by comparison with the implications of a decree of 1556. This imposed on every hereditary landowner the duty of providing one fully armed and horsed warrior for every hundred cheti of cultivable land in his possession. The full significance of this decree emerges only when taken in conjunction with the boyar's loss of the right to leave the service of one prince for that of another and still retain his land. Not only could he not leave Muscovy without committing treason: after 1556, he was obliged to serve the Tsar of Muscovy, on pain of losing his land. He no longer enjoyed his landholding as a source of power in itself. The boyar became a conditional landowner, dependent for his continued tenure on the fulfilment of certain conditions. In other words, he was coming close to assimilation with the lesser gentry, the *pomeshchiki*, who also held their land on terms of service. The distinction was further diminished in 1562 and 1572 when further decrees made it almost impossible for a boyar to dispose of his land by any means or for any reason at all. Ivan even limited the right of the boyar to bequeath his estate at will. The land must not go out of service. It had always been a principle of Muscovite policy that he who serves must have land. To this was now added the further principle that he who has land must also serve.

The importance of this new principle can scarcely be over-estimated; it made of Muscovy a compulsory service state in which each class was bound by ties of service, albeit sometimes indirect, to the autocracy. Not until the second half of the eighteenth century was the nobility able to enforce some loosening of these onerous bonds.

Two political theories heralded and justified ideologically Ivan's changing system. The first concerned the autocracy; the second the service nobility. Ivan was his own ideologue and, in his famous and bitterly polemical correspondence with Prince Kurbsky,[1] he outlined a theory of personal autocracy. Had the Russian autocrats ever shared their power? he asked. No, Ivan thundered, 'they themselves from the beginning have ruled all the dominions – not the boyars, and not the grandees.' He then asserted the utility of individual power as practical politics. 'And with your lawless eyes [i.e. the boyars'] you yourselves have seen what happened as a result of this in Russia, when in each city you were governors and viceroys, and what destruction was wrought as a result of this.' Finally, the head-piece of Ivan's conviction of the divinity of his power – expressed in even stronger terms than those used by Abbot Joseph of Volokolamsk: 'Did I,' Ivan asked Kurbsky, 'ascend the throne by robbery or armed force of blood? I was born to rule by the grace of God.'

The same point of view, but from the angle of the service nobility, was put forward by a certain Ivan Peresvetov in a number of epistles to the Tsar. Peresvetov's ideal was a strong autocrat and a centralized state, both based on a class of de-voted servants, functioning either as the Tsar's soldiers or offi-cials. There must be an end to boyaral privileges and to status that was not derived from merit. The boyars fought only for their own selfish aims. They must give way to a new class that would be identified with the State. Only in this way could the

1. For Kurbsky, see p. 53.

Tsar fulfil his true duty – to bring about the reign of justice in Muscovy.

This dispute between the service nobility and the boyars corresponded to deep-seated changes which had been gathering force for the best part of a century. They all tended to undermine the economic power of the magnates.

Agriculture and rural pursuits were, as ever, the overwhelmingly predominant branch of the national economy. But inside this framework, forces were at work that would in time drastically transform the closed agricultural system of early Muscovy. Foremost among these was the growth of mercantile capital. In 1466, for example, a certain Nikitin, a merchant of Tver, travelled to Azerbaijan, Persia, and India. The fact is that the consolidation of Muscovy as a national state had, by sweeping away many internal customs barriers and obstacles to trade, created a vast national market. This in its turn encouraged an increase in commodity production. Moscow, for example, one of the largest contemporary European cities, had a population of some 200,000 in the 1560s. Foreign observers counted a daily total of 700–800 cartloads of grain entering the city's gates. This is a clear, perhaps extreme, example of the breakdown of local self-sufficiency, but it is nevertheless significant.

A large part of this trade was conducted outside private hands. The State itself had a quasi-monopoly of such goods as furs, timber, and grain, and numerous monasteries supplied salt, fish, and butter over a wide area. The most efficiently worked estates belonged in fact to ecclesiastical institutions.

Muscovy's landlocked position severely limited foreign trade, of course, but there was still interchange with the tribes beyond the Volga, with Genoese colonies in the southern Crimea, and with Asia Minor as far to the south as Egypt and Byzantium.

These possibilities, both inside and outside Muscovy, made their impact at the village and municipal level. The manor remained, on the whole, the unit of production, and money was still rarely used; but it was here that a gradual division of labour took place. Specialized 'non-ploughing' peasants were removed from the land and employed in the manufacture of cloth, farming implements, baskets, twine-weaving, and so on.

Such craftsmen might even occupy special villages or inhabit the new trading quarters – the *possad* – which were growing up as a commercial adjunct to the towns. The same development also accounts for the increase in the number of fairs and market days. These were media for exchanging commodities produced in excess of local needs. But perhaps nothing was more indicative of the slow transition to a money economy in the countryside than the partial disappearance of *obrok* – deliveries in kind – made by the peasant to his lord, in favour either of a money payment or of increased *barshchina* – labour by the peasant on the lord's land. It had by now become much more profitable to secure a share of the peasant's labour and develop one's own estate rather than receive some eggs or mutton or flax.

A nation-wide market was growing up, with a production of surplus commodities, a developing system of labour specialization, transition to money as a medium of exchange – and in this new world, the closed, self-sufficient manorial system of the boyar magnates was an obvious anachronism. There is an instructive contrast here with Western Europe, where the manor had long since ceased to be an economic unit and had given way to developing capitalism and to advanced credit and banking institutions. But in Muscovy the contemporaneous decline of the land-owning class was the economic reality that made sense of, and justified, Ivan IV's policy of squeezing his boyars out of their hitherto dominant posts in the army, the administration, and the rudimentary system of government.

The other side to this coin was Ivan's powerful advancement of the lesser serving gentry. The origin of these *pomeshchiki* goes back to the second half of the fifteenth century.[1] This practice continued on a growing scale until, by the mid sixteenth century, the *pomeshchiki* were the mainstay of the autocracy and the new Muscovy emerging from the turmoil of the almost incessant warfare on its western, eastern, and southern frontiers.

Indeed, it was war that conditioned the growth of the *pomeshchiki*. So long as Muscovy's wars were waged against

1. See pp. 32 and 37.

the other Slav principalities, no need arose for large semi-permanent military forces. But when the enemy became the more highly developed Poles or Swedes or Lithuanians, or it was a question of defence against the raids of the Crimean Tartars, the old haphazard system of army organization no longer sufficed. There were longer borders to be defended, and more skilful and better-armed enemies to contend with.

This was the background to the wholesale redistribution of land throughout the first half of the sixteenth century in much of Muscovy. Every category of land was thrown into a vast pool to satisfy the State's demands – palace lands belonging to the Tsar, 'black lands' occupied by free peasant communes, and the new lands conquered to the south and east. The pressure was so great – in a Muscovy, of all realms, where land had hitherto always been abundant – that even the sacrosanctity of Church lands was put in doubt. The question of monasterial landholdings coincided, not accidentally, with much debate inside the Church concerning the morality of such sources of wealth. The ascetic influence of Mount Athos was prominent in the case of those monks, such as Nil Sorski, who urged the Church to divest itself of its properties. But although he and his supporters were backed, from outside, by Ivan III and Ivan IV, both of whom would gladly have secularized the estates of the Church, no substantial inroad was made. All that Ivan the Terrible could do was to enact a decree in 1580 that forbade further monasterial acquisition of boyaral lands, and this slowed down considerably the growth in monasterial estates.

Who were the *pomeshchiki*? Their social composition was heterogeneous and even included Lithuanians and Mongols. As for the Slav elements, they were by no means drawn from the original Muscovy alone, but could and did hail from any of the former independent principalities conquered by Muscovy on its path to predominance. The core of the new class was formed of those boyars and princes, together with their military servitors, who had lost their original function on their incorporation into Muscovy. But the *pomeshchiki* of this type were vastly outnumbered by very diverse strata of the Muscovite population: merchants, palace servants, townspeople, clerks, even slaves, if they had been accustomed to bearing arms in their master's service, all these became the backbone of the new Muscovy of Ivan the Terrible.

They all received a grant of land (and at times a small, irregular salary) in return for a form of state service that was usually military in character.

Very soon other measures were taken to reinforce the position of the *pomeshchiki*. They were entitled to judge the peasants on their land; they collected taxes on behalf of the State; they were removed from the control of the provincial governors and made directly answerable only to the central power; most important of all, they were allowed to bequeath their holdings on condition that their heirs inherited their service obligations.

In all this there was in the making an implicit alliance of incalculable importance to all Russian history until the middle of the nineteenth century. At the centre the Tsar would enjoy autocratic power; locally, his authority and policy would be exercised through the landowner. By and large this pattern remained unchanged and became central to the structure of the Russian State and Government.

In the 1550s this implicit alliance could not be finally sealed until the peasant question was solved in a manner favourable to the *pomeshchiki*; in other words, until the latter had a ready source of labour to work their land. Thus the ascent of the serving gentry was matched lower down the the social hierarchy by a tighter control over the primary producer, the peasant. What use was land without the hands to work it?

In the late fifteenth and early sixteenth centuries many peasants were already beginning to lose their freedom through the mere fact of having lived and worked on the same land for several decades. There were the so-called 'old-dwellers' whom custom had gradually deprived of the right to leave the estates where they had been settled for generations. But what finally converted the 'old-dwellers' and other peasants into serfs was not only custom but also the economic and military requirements of the State. Serfdom took root as a complex blend of custom, economic indebtedness, peasant poverty, legal enactment, and *raison d'État* – all within the context of the type of encompassing autocratic State that was fast developing.

Legal limitations on the peasant's freedom of movement were the first step towards complete loss of mobility. In the fifteenth century, in Pskov, for example, there already existed a judiciary charter which limited the peasant's right of departure to Philip's Feast Day, 14 November. More significant, however, was Ivan III's Code of 1497. This inaugurated a series of decrees, extending over the next century and a half, all of which were expressly intended to deploy the power of the State on behalf of the *pomeshchiki*. The Code of Ivan III, while confirming the right of departure, stipulated that the peasant could exercise this right only by giving due notice and fulfilling all his outstanding obligations a week before or after St George's Day, 26 November, the Russian equivalent of Michaelmas. By then the harvest would have been gathered in.

In 1550 this limitation was repeated in Ivan IV's Code. But, towards the end of the century, Ivan's foreign wars and financial exactions produced an incipient financial crisis to which the ever more frequent response was flight – generally to the open lands of the south. This in its turn led to an intensification of the State's pressure. In 1597 Ivan's successor, Boris Godunov, ordered all landowners to return to their previous owners all those runaway peasants who had fled during the five years preceding – in other words, since 1592. By implication this made the landowner the proprietor of those peasants who had not fled. But two factors delayed the final appearance of serfdom: first, the interest of the larger landowners in securing an adequate labour supply, if necessary by kidnapping and bribery; second, the famine and upheaval of the Time of Troubles in the early seventeenth century, which frequently forced even the landowners to abandon their land. But with the return to stability, the gentry began to press the new Romanov Tsar with petitions urging the complete abolition of the period of redemption; and in 1642 this was extended to ten years, although not before many peasants had left gentry-owned land for that of the 'powerful people'. In 1646 'registration books' recorded the relationship of all dependent agricultural workers to the landowner, and in 1649 the 'Assembly of the Land'[1] confirmed that the landowner became the absolute and unconditional master of his peasants, as well as the tax-gatherer on the Tsar's behalf. This was the head-piece set on the edifice of serfdom, whose foundations, legal and economic, had been laid more than a century and a half earlier.

Ivan the Terrible died long before the ultimate stage in this process was consummated. But his own policy and the trend of economic development did much to bring it about. To him it was vastly more momentous to encourage a class of dependent servitors whose backing he could secure in his struggle against the boyars. Indeed, he vindicated the policy of his father and grandfather. All the more significant economic developments of the first half of the sixteenth century – the development of handicrafts, the growth of the market, the advance to the three-field system in agriculture – favoured the service nobility far more than the boyars, burdened as these latter were with the maintenance of seigniorial magnificence, the upkeep of private armies, and intermittent strife with their fellow magnates.[2] The *pomeshchiki* had no such difficulties to contend with, and could draw a proportionately higher benefit from the over-all economic advance. Ivan's policy was abundantly justified.

THE PERIOD OF OPRICHNINA

The shift in power that the elimination of the boyars involved was no peaceful sequel to economic change. As early as 1553, a revealing episode showed that the boyars had by no means resigned themselves to the Tsar's policy. In that year Ivan fell ill and, fearing he would die, called on the princes and boyars to swear allegiance to his infant son, Dmitry. This would have entailed acknowledging as Regent the Tsarina Anastasya, whose lowly birth had already provoked hostile comment. Many of

1. See p. 63.
2. The story is even told that one of Moscow's leading boyars was so impoverished by his military exertions on behalf of Ivan that his wife had nothing to wear for her daughter's wedding – the husband had pawned his wife's whole wardrobe.

the boyars refused to kiss the cross. In this they were joined by Ivan's two closest advisers, Silvester and Adashev. The opposition planned to exclude Dmitry from the succession and to nominate their own ruler, a cousin of Ivan's. Ivan could not even be confident that his wife and son would long survive his own apparently imminent death. He soon afterwards recovered, in fact, but he did not forget.

The atmosphere was further thickened and embittered in 1560, when Anastasya died. Accusations of having caused her death by witchcraft were made against Silvester and Adashev, who were in any case already beginning to fall from favour, and although the epoch of true terror had not yet set in, the two advisers and some of their followers were exiled or imprisoned for life. Ivan himself suffered a fearful blow in the death of his wife, and gave himself up to the most obscene and blasphemous orgies. Four further marriages and two long periods of concubinage followed.

It was, however, war and military policy that brought to a bloody head Ivan's conflict with his boyars and advisers. After the victories at Kazan and Astrakhan the question was, where would Muscovy move next? To the south, against the Crimean Tartars; or to the north-west, against the Livonian Knights; or to the west, against Poland and Lithuania? The boyars and Ivan's closest advisers would have preferred a campaign against the Tartars, but preliminary reconnaissance in 1555 and 1556 soon revealed to Ivan the absurdity of a war waged across the steppes. Besides, behind the Tartars stood the Turkish empire of Suleiman the Magnificent. Livonia, by contrast, was relatively near at hand. It also offered access to the Baltic with the great ports of Narva, Revel, and Riga. The shores of the Baltic were of silver and its waters of gold, said Ivan. To backward Russia, with a rudimentary agriculture and a virtually non-existent industry and technology, the Baltic offered the prospect of benefit from the more advanced countries of the West. The West offered power.

The importance of Ivan's move is dramatically reflected in a letter sent by King Sigismond Augustus of Poland to 'our dear sister and kinswoman', Queen Elizabeth of England. At this time, 1559, English seamen and merchants were already using Narva as a base for Anglo-Russian trade.

... Now we write again to your Majesty [complained Augustus] that we know and feel of a surety, the Moscovite, enemy to all liberty under the heavens, daily to grow mightier by the increase of such things as be brought to the Narve, while not only wares but also weapons heretofore unknown to him, and artificers and arts be brought unto him: by means whereof he maketh himself strong to vanquish all others. ... We seemed hitherto to vanquish him only in this, that he was rude of arts and ignorant of policies. If so be that this navigation to the Narve continue, what shall be unknown to him? ... For now we do foresee, except other princes take this admonition, the Moscovite puffed up in pride with those things that be brought to the Narve, and made more perfect in warlike affairs with engines of war and ships, will make assault this way on Christendom, to slay or make bound all that shall withstand him: which God defend.

The fate of the Slitte mission is also testimony to the far-sightedness of Ivan's policy and the consequent fears it aroused in central Europe. Hans Slitte was a German from Goslar, something of an adventurer, who went to Moscow in 1547. He learnt Russian and became an intimate of the Tsar, with whom he would often talk of German progress in the sciences and arts. Ivan then gave Slitte the task of recruiting for service in Muscovy specialists of all kinds – engineers, architects, artisans, doctors, chemists, printers, and even theologians. Fortified by the permission of the Holy Roman Emperor, Charles V, Slitte assembled more than a hundred German specialists. And there the enterprise ended: the Hansa towns were no more willing than Augustus to see a Westernized Russia. Slitte was arrested in Lübeck at the order of the town council, and his followers dispersed.

On the trumped-up pretext that the Bishop of Dorpat was in arrears with a long-forgotten tribute, Ivan declared war on the Livonian Knights in 1558. At this time, their territory embraced Estonia, Courland, and Livonia. The position of the Knights was already in jeopardy through the progress of the Protestant Reformation. Ivan, by dint of superior numbers, had soon

taken Narva and Dorpat and some other fortified places in the
north of Latvia. But when, three years later, the territories of
the Knights were taken over by Poland-Lithuania, Denmark,
and Sweden, Ivan found himself faced with impossible odds.

The war dragged on for the remainder of Ivan's reign, strain-
ing Muscovy to the utmost. Dissension at home, among the
boyars, was all too manifest in their occasional attempts to flee
westwards. Setbacks at the front combined with a thickening of
the hostile atmosphere to persuade many of the leading boyars
– a Glinsky or a Belsky, for example – that they had better go
while the going was good.

It was Prince Andrew Kurbsky who was responsible for
Ivan's final onslaught on the boyars and brought to bursting-
point all the internal tension accumulating since the early days
of the reign. Kurbsky, the outstanding Muscovite general and
a friend and adviser of Ivan, lost a battle against the Poles,
despite his vastly superior numerical strength. Soon afterwards
another boyar general, Prince Shuisky, was also defeated. Quit-
ting his headquarters in Dorpat, Kurbsky fled to Lithuania,
where he took up arms against his former ruler, and thus precipitated one of the bloodiest epochs in Russian history.

Ivan first wrought his vengeance on Kurbsky's family, killing his wife and son. At the turn of the year, a simulated abdi-
cation and withdrawal from Moscow to Alexandrovskaya
Sloboda, a wooded district a short distance to the north of the
capital, brought him unlimited power from the Moscow populace to rule as he alone thought fit. According to a contemporary chronicler, the people of Moscow were so thunderstruck
at the thought of losing their Tsar that they compared themselves to sheep bereft of their shepherd.

When Ivan returned to the Kremlin, sworn to extirpate all
villains and traitors, he carried out an extraordinary innova-
tion: he divided Muscovy into two distinct and independent
realms – an Oprichnina and a Zemschchina. The former term
had originally signified land set aside for a widow's permanent
tenure; the latter term is connected with 'Zemlya' – 'land'. Ivan
revolutionized the meaning of both. The Oprichnina was that
part of Muscovy to come under his personal rule. The adminis-

tration of Zemschchina he left to a council of boyars, Ivan reserving to himself only a formal sovereignty. For a time, a converted Tartar was even installed as Tsar of the Zemschchina. Ivan also established a separate Court, with all its appurtenances, in Alexandrovskaya Sloboda.

This division of his country at once inaugurated Ivan's reign of terror. Sadism had always been a marked feature of his character. But the excesses and outrages of the Oprichnina period clearly show a mind unhinged. He inaugurated government by pure terror. This is not always clear, for apart from the scope of the destruction, Ivan tried to conceal his policy from foreign observers. But some idea of his motives can be derived from an analysis of the territorial composition of the Oprichnina. It takes shape gradually as a wedge, with its base lying on the northern towns and villages, cutting through the central area of Muscovy, splitting the Zemschchina in two. In the central area the two territories were intermingled. To the Zemschchina were left the Perm and Vyatka towns and Ryazan on the east; Pskov and Novgorod on the west; and Smolensk on the south-west. At its greatest extension, the Oprichnina covered not only half the country but comprised all the wealthier towns, trading routes, and cultivated areas; equally important, it also covered the areas where the old boyaral lands predominated.

These were Ivan's target. Kurbsky recognized this when he wrote in his history of the Grand Prince of Muscovy: 'They [that is, the boyars] had great *votchini*; I think probably for that he destroyed them.' ('*Votchina*' was the technical term for a boyar's estates.) At first Ivan's persecution had not its later scope. It was limited to the associates of Kurbsky, Silvester, and Adashev, those who had opposed the succession of Dmitry in 1553, and those suspected of preparing to flee westwards. Ivan tortured them, executed them, or sent them into exile.

The Oprichnina later degenerated into terror almost for the sake of terror. Its main instrument was a select corps of 1,000 Oprichniki, later increased to 6,000. who were uniformed in black. Their badges were a dog's head and a broom – symbols of their doglike devotion to the Tsar and their duty to sweep away treason from the Muscovite State. The atrocities wrought by the Oprichniki reached their peak in the sacking of Novgorod in 1570. The town was suspected of seeking union with Poland. The precise facts of the matter are unknown. The mere suspicion, however, provoked an onslaught by the Oprichniki in which more than 60,000 people were killed and the town and its surroundings devastated. The slaughter here and elsewhere was relieved only by Ivan's macabre donations to convents so that masses might be offered up for the souls of his victims.

The killings and destruction are the most obvious features of the Oprichnina, but they cannot obscure its political meaning. As far as can be judged, it was primarily an instrument for physically uprooting and destroying the class of boyars and princes. Both in the Oprichnina and the Zemschchina they were forcibly dispossessed and evicted, and their estates parcelled out among Ivan's serving-men as *pomestiya*. The 'chief nobility', to quote Giles Fletcher, English Ambassador to Ivan's son and successor, 'were equalled with the rest'. Wrenched from their local roots and followers, the boyars, Ivan hoped, would no longer be able to exert any particular influence. He would have created a state in which no local power stood between the people and the autocrat. In actual fact, the years after Ivan's reign saw a partial boyar counter-revolution. But he had, for all that, created the foundation for the Tsardom of the seventeenth century; and not only the seventeenth century – that of the eighteenth and nineteenth centuries also. He had produced an amalgamation of the State and the landowners to which the only threat would come from the peasants and the intelligentsia.

The upheaval of the Oprichnina naturally had unfavourable repercussions on the Livonian war. It was directly connected with the defeatism that afflicted large numbers of the Muscovite forces. To that unhappy campaign was soon allied the great Tartar attack on Moscow in 1571. Devlet-Girei, the Crimean Khan, aided probably by disaffected boyars, took advantage of Ivan's external and internal difficulties to move north with

120,000 men. Ivan dared not even defend his capital, after so many reverses elsewhere. He fled to Yaroslavl. The Tartars set the city's suburbs on fire. The flames spread to the main city, where thousands of refugees had taken shelter. Much of the population perished in the holocaust, and the Tartars dragged off many more thousands to slavery.

On the Livonian front the invasion forced Ivan to limit himself still further to holding operations. Another decade of war brought no lasting success, and in 1582 he had to admit complete and utter defeat. He had to renounce all his conquests in Livonia. The Baltic was as remote as ever.

This failure was partially mitigated by two unplanned and unexpected successes – contact with England and the conquest of Siberia. In 1553 the *Bona Fortuna*, commanded by Richard Chancellor, the sole surviving vessel of three which had originally set sail in the quest for the North-East Passage, arrived at the mouth of the Dvina. Chancellor travelled on to Kholmogory and thence to Moscow. Here he was warmly welcomed by Ivan, only too willing to relieve his dependence on German traders. This was the prelude to the formation of the 'Russia Company'. Ivan's charter of 1556 granted the Company extensive trading privileges, which included the right to customs-free trade and tax exemption for English trading counters established in Kholmogory, Vologda, and Moscow.

The reactions of the English traders were on the whole hostile and certainly condescending. They found Muscovy ruled by an all-powerful Tsar, they found the Russians drunkards and hypocritical, the Orthodox Church rife with superstition and with morals perverted by homosexuality. George Turberville, a poet, and secretary to one of the English embassies, penned in verse a number of such criticisms:

> A people passing rude, to vices vile inclined,
> Folk fit to be of Bacchus train, so quaffing is their kind.

> Devoutly down they duck, with forehead to the ground,
> Was never more deceit in rags, and greasy garments
>   found.

Turberville even compared the Russians to the Irish:

Wilde Irish are as civil as the Russies in their kind,
Hard choice which is the best of both, each bloody,
    rude and blind.

Chancellor, on the other hand, found matter for praise in the
leniency of Russian laws. In Elizabethan England, a man was
hanged for stealing; but in Russia he was only tortured, and
was not hanged until he had committed his third offence.

Be that as it may, trade was extremely profitable, producing
net profits of several hundred per cent. Soon English ships were
making an annual call at Kholmogory – and later at Archangel,
which quickly developed from a small fishing village into a
port of some importance. It suffered from the disadvantage of
being ice-bound for some eight or nine months of the year.
Even so, it was Muscovy's only direct sea-route to the West and
was thus indispensable to Ivan, all the more so when success
in Livonia seemed more and more remote.

The traders who used the Archangel route were not only
English; there were also Dutch and French. But the English
were by far the most numerous and enterprising, and the most
favoured (although Ivan's suspicions caused difficulties at
times). There were English warehouses, trading counters, and
agencies in a number of Russian towns – Yaroslav, Novgorod,
Kazan, Astrakhan. In Vologda and elsewhere there were
English factories for the manufacture of rope and twine. In
Pskov, Nizhni Novgorod, and Moscow, English coins circulated.

Muscovy was certainly the junior partner in its trade with
England. It supplied primarily natural products and raw
materials in return for manufactured goods – pelts, skins, wax,
tar, pitch, ships' masts, hemp, flax, were bartered for cloth,
arms and ammunition (despite Polish and Swedish protests),
silks, spices, lead, and copper. Nothing is more indicative of
Ivan's awareness of the backward state of Muscovy than the
encouragement he gave to English artisans and specialized
workers to settle in Muscovy. They included medical men, a
midwife, shipbuilders, a goldsmith, and an apothecary.

This trade had very strong political and also marital overtones. In Ivan's correspondence with Queen Elizabeth, there was talk of a marriage to Lady Mary Hastings – a relative of the Queen – 'which entreaty', in the words of the English Ambassador, Sir Jerome Bowes, 'by means of her inability of body, by occasion of much sickness, or perhaps, of no great liking either of herself or friends, or both, took no place'. Ivan's chief political aim – an Anglo-Muscovite offensive–defensive alliance against Poland and Sweden – also 'took no place'. Yet his relations with England and the Queen always remained sufficiently intimate for Bowes to be told when news came of Ivan's death: 'Your English Tsar is dead.'

The Muscovite conquest of Siberia was the other partial compensation for Ivan's failure to reach the Baltic. As early as the fourteenth century, Novgorod merchants had ventured into western Siberia beyond the Urals. But not until after the capture of Kazan did the movement seriously get under way. It then went rapidly ahead, proceeding from the north. The only opponents were the primitive tribes, armed with nothing more than bows and arrows. In 1558 Ivan made a large grant of land along the Kama River to the Stroganov family of merchant-adventurers and colonizers. This settlement was soon developed into a vast commercial enterprise which exploited to the full the natural resources of the region and also of the trans-Ural areas – iron-ore deposits, copper, tin, lead, and sulphur. Muscovite penetration has been well compared to the operations of the East India Company in combination with the Spanish conquest of Peru and Mexico. In 1582 the process went a considerable step further when the Tartar State of Kuchum, the only force offering any serious resistance to the Russians, was overcome by a Muscovite Cossack army of less than a thousand men from the Don and Volga. Yermak, the Cossack leader in the service of the Stroganovs, advanced from the Kama to the Irtysh and took the Tartar capital of Sibir, the whole territory of which was then annexed to Muscovy. Regular troops and governors were now sent eastward from Moscow, and towns and strong-points set up in the newly conquered areas. Tyumen was founded in 1586, Tobolsk in 1587, Berezov in 1593, Narym in 1596. Ivan did not see all these later developments. But in his last years he was undoubtedly comforted by the thought that his first governors were entering on a new and rich inheritance.

This, on the other hand, could by no means make up for the remorse he felt at the death of his eldest son – whom he himself had killed in a blaze of anger in 1582. When Ivan's own health began to fail he called in the aid of witches and soothsayers from the north. Sir Jerome Horsey, an agent of the Russia Company, has left a vivid description of the Tsar's last days:

The Emperor began grievously to swell in his cods, with which he had most horribly offended above 50 years together, boasting of a thousand virgins he had deflowered and thousands of children of his begetting destroyed.

Every day the Tsar would be carried into his treasure-chamber, where he would fondle and discourse on the curative powers of precious stones – turquoise, coral, diamond, ruby, emerald, sapphire, onyx. But none could help. The verdict of the soothsayers was at one with that of some spiders used as a means of divination. The end came while Ivan was preparing to play a game of chess:

He sets his men [1] ... The Emperor in his loose gown, shirt and linen hose, faints and falls backward. Great outcry and stir; one sent for aqua vita, another to the apothecary for 'marigold and rose water' and to call his ghostly father and the physicians. In the mean he was strangled and stark dead.

Special guards were thrown round the Kremlin in case of disorder. But the succession passed easily enough to the Tsarevich, Theodore Ivanovich. This was in fact wholly deceptive. Ivan's attempted social revolution and his last twenty-five years of war had brought Muscovy to the verge of anarchy. The true, prophetic verdict is that of Fletcher, the English Ambassador to the new Tsar:

This desperate state of things at home maketh the people for the most part to wish for some foreign invasion, which they suppose to be the only means to rid them of the heavy yoke of his tyrannous government.

1. 'All saving the king, which by no means he could not make stand in his place with all the rest upon the plain board.' (Horsey's footnote.)

CHAPTER 5

The Time of Troubles

THE Moscow branch of the dynasty of Rurik came to an end in the person of Ivan the Terrible's younger son, Theodore. A line of strong Tsars reached a sudden nadir when a physically degenerate and mentally enfeebled autocrat, altogether unfitted for his place in the Muscovite scheme of things, ascended the throne of Muscovy. The new Tsar, crowned in 1584 at the age of twenty-seven, amid the full panoply of the traditional ceremony, had a perpetual simper playing about his mouth. He was of devout character; his favourite pursuit was bell-ringing. But he also enjoyed the antics of dwarfs and court jesters, and combats between men and bears entertained him.

The hazard that brought such an incapable Tsar to the throne was all the more fateful in that his reign coincided with the burgeoning of the problems inherent in his father's policies. 'God hath a great plague in store for this people,' wrote Sir Jerome Horsey prophetically.

In the absence of a strong Tsar, there was a recrudescence of internecine boyaral strife around the throne and of boyaral efforts to win back the privileges lost during the previous reign. The peasantry suffered most from the high taxes resulting from the Livonian wars, and were beginning to flee financial oppression, loss of personal liberty, and unending military levies. The central areas, the heart of Muscovy, became depopulated, as villagers streamed to the 'black earth' territory of the Don, the upper Donetz, and the middle Volga. Towards the turn of the century, Fletcher noticed fifty large but completely deserted villages along the stretch of some ninety-odd miles between Vologda and Yaroslav. This was only one of many such examples.

This loss of manpower was one of the most potent factors in the economic crisis that followed. Flight was not the only expression of peasant discontent. For the first time in Muscovite history, the early years of the seventeenth century were marked by large-scale peasant revolts – elemental and chaotic in organization and aims, but no longer local or regional. Given this disarray inside a weakened country, there was ample leverage for foreign intervention, of which the Poles and Swedes were not slow to take advantage. In fact, the 'Time of Troubles' – the period between the death of the Tsar Theodore Ivanovich in 1598 and the coronation of the first of the Romanov dynasty in 1613 – is often known as the period of peasant wars and foreign intervention. Class rose against class. Tsar followed Tsar, dubious pretenders gathered support as best they could. Poles and Swedes occupied important areas of Muscovite territory. In those fifteen years, Muscovy came near to collapse in utter anarchy.

Theodore was so patently unfit to govern that recourse was had to a Regent, the first being Nikitin Romanov, a maternal uncle of the Tsar. But he died in 1586. The enigmatic figure of Boris Godunov, the hero of Pushkin's poem-drama and Mussorgsky's opera, took his place. Godunov hailed from a family of Mongol origin that had first taken service with the Grand Prince of Moscow in the fourteenth century. Under the Oprichnina he rose to rapid eminence, and succeeded in marrying his sister Irene to Theodore, thus making himself the new Tsar's brother-in-law. This upstart rise provoked envy and hatred among the old boyar families; and in the early years of Godunov's regency, there were several plots against his life as well as attempts to alienate him from his proximity to the throne by trying to bring about a divorce between the Tsar and his childless consort. These attempts failed, and their instigators – the boyar families of Mstislavsky, Shuisky, Belsky – were executed, tortured, or exiled. A number of merchants were also publicly beheaded. Godunov further reinforced his position by installing his own nominee, Job, the Archbishop of Rostov, as Metropolitan. A year or two later, Godunov, exploiting the Orthodox Church's aversion to subordination to the Patriarch in Constantinople, a subject of the infidel Sultan, prevailed upon Constantinople to create a Patriarchate of Moscow and All-Russia. This ranked after Constantinople, Alexandria, Antioch, and Jerusalem.

In 1587 Godunov had disposed of or silenced all his rivals. He now ruled supreme, even though he was still nominally Regent. He was quite illiterate, but this did not prevent him from directing all correspondence with the heads of foreign states. Godunov received ambassadors in his own palace with as much pomp and ceremony as in the Kremlin itself.

In many ways Godunov's rule followed the pattern, at home and abroad, set by Ivan the Terrible. There was the same hostility to the boyars and the same reliance on the lesser serving nobility. In particular, Godunov intervened strongly on the latter's side in the peasant question. In 1597 he issued a decree entitling a landlord to retake by force any peasant who had fled during the past five years.[1] This affected the larger landowners, who were always in a stronger position to attract labour through their ability to offer less difficult and exacting conditions of employment. Later decrees intensified the effect of the first by expressly forbidding the larger estates to attract peasants from the smaller.

Abroad also, Godunov pursued Ivan's aggressive policy, though less forcefully. Smolensk and Astrakhan were fortified with stone walls. In the steppe zone, fortress towns such as Kursk, Voronezh, and Byelgorod were established. Godunov even succeeded in recapturing from the Swedes some of the areas on the Gulf of Finland yielded up by Ivan – Ivangorod, Yam, and Karelia.

But these were the years before the storm. Apart from Theodore, Ivan had had one other son – Dmitry, the offspring of his marriage with Maria Nagaya. Early in the reign of his halfbrother, Dmitry and his mother had been exiled to Uglitch, lest they serve as a tool for disaffected boyars anxious to unseat Godunov. But in 1591, the nine-year-old boy was found murdered, his throat cut by a gang of hired killers. The instigator of the crime was never discovered. But the rumour soon gained widespread credence that Godunov, anxious to eliminate the

1. See p. 49.

last remaining male heir to the throne, had engineered the murder.

Be that as it may, the death of the Tsar Theodore in 1598 further thickened the atmosphere. Power passed to the Tsarina Irene, but she soon abdicated and took the veil. There was no choice but to tread the Polish path and *choose* a Tsar – an unprecedented situation in Muscovite history. Godunov, by virtue of the power that he had in fact exercised, was the favoured candidate. But before he would accept the throne, he insisted upon election by the Zemsky Sobor – Assembly of the Land – a rough equivalent to the Estates-General of Western Europe. This body had been called into existence during the sixteenth century. It had never developed or displayed any independent viewpoint. It was used in the main to endorse a policy that the Tsar had already decided on. When it met in 1598 it consisted of 474 representatives drawn from the clergy, the boyars, the lesser nobility, officials, and merchants. Godunov's agents were active among the Sobor so that its choice was unanimous. Even so, Godunov insisted on repeated pleas before he would accept the throne. Given the abundant evidence of boyar opposition and his reputed murder of the legitimate heir to the throne, the new Tsar had to ensure that his election appear but a reluctant response to the demands of the Sobor.

In this Godunov succeeded, and, until famine nullified his efforts, his reign looked promising. He was, for example, the first Tsar to take an interest in education, and he even toyed with the idea of establishing a university, and sent a number of young Russians to study abroad in the West. Four came to England. For about twelve years Muscovy lost track of them. The only one who could then be traced was found to be earning his living as a minister of the Anglican Church. Godunov also tried to eliminate the more flagrant abuses in the workings of the judicial machinery. But all this prospect of better things was soon destroyed by the conflict with the great boyar families, who had never reconciled themselves to Godunov's 'election' as Tsar. Not for nothing did Godunov, for fear of poison, maintain six foreign doctors at his court.

One of the first to suffer was Bogdan Belsky, dispossessed and then exiled to a distant region of the Volga Basin. Members of the Shuisky family were dispatched as viceroys to outlying provinces. Godunov reserved his most brutal measures for the Romanov family, from whose ranks had come Anastasya, Ivan the Terrible's first and most popular wife. Trumped-up accusations of sedition and poisoning attempts were followed by tortures, and deportation to the north of Siberia and to remote monasteries. In the end, Godunov's campaign against the boyars degenerated into a welter of denunciation and torture, reminiscent of the excesses of the Oprichnina.

A succession of disastrously poor harvests from 1601 to 1603 completed the social isolation of Godunov's régime. The basis of Muscovite subsistence, and especially of Moscow and the surrounding districts, rapidly succumbed to the resulting speculation in grain, and to the depopulation and earlier agrarian collapse of the central provinces. Whole villages died of starvation. Men were reduced to eating grass and birch bark; they took to cannibalism. Many landowners, unable to feed their peasants, drove them from their estates; or peasants would voluntarily engage themselves as slaves, for the mere promise of food. One immediate consequence was the formation of marauding bands of homeless peasants, their sole aim to attack and raid the granaries of the *pomeshchiki* and the merchants. In 1603, one such band almost reached Moscow before it was bloodily suppressed by regular troops.

Godunov faced this catastrophe helplessly and hopelessly. Given the inflated price of grain, the Government, it has been calculated, disbursed to the starving populace only about a third of the sum needed for subsistence. Moreover, he concentrated in the cities such famine relief as there was. In Moscow, for example, the announcement of free grain supplies attracted an unmanageable throng of the needy, thereby intensifying the crisis still further.

It was at this moment of governmental collapse that Polish intervention began to play its part. Its instrument was the first false Dmitry, and its allies the boyars. As early as 1598, when the election of a successor to the feeble-minded Theodore was in train, there had been rumours that either Godunov or the

Romanovs would put forward a pretender to the throne in order to sabotage the election of the other. The obvious choice was Dmitry, the nine-year-old son of Ivan the Terrible, murdered at Uglitch in 1591. The circumstances of his death were so ambiguous that belief in his survival was not altogether inconceivable. Besides, as Korolenko, the neo-Populist writer of the late nineteenth century pointed out, Russia is characteristically the land of pretenders. In Western Europe the same yearning for a hidden Messiah produced the false Manfreds and false Friedrichs of the Hohenstauffen era. In Russia, in the seventeenth and eighteenth centuries, it produced hundreds of false Alexeys, false Peter IIs, and false Peter IIIs. Each would unveil his true identity at a pre-ordained time and rescue the people from their misery.

All the same, the first false Dmitry had sound power-political backing. His true origins have never been uncovered. But, as Klyuchevsky puts it, there is no doubt that he was cooked in a Polish oven. Soon after the turn of the century, rumours of the reappearance of Ivan the Terrible's son began to circulate in Poland and Lithuania. In 1603 the rumours took on tangible form in the shape of a young man, claiming to be the Tsarevich Dmitry, who appeared, with some evidence of authenticity, at the Ukrainian castle of the Polish aristocrat, Adam Wisniewicki. Wisniewicki believed his visitor's story and introduced Dmitry to others of the Polish nobility, in particular to the heavily indebted George Mniszek, *voevoda* of Sandomir, who saw in the prospect of backing the putative Tsar of Muscovy a means to saving his family fortunes. He also affianced his daughter Marina to the Pretender, on condition that her suitor made himself Tsar within a year. The Papal Nuncio in Poland was also interested. He saw in the Pretender, who did in fact secretly leave the Orthodox Church for Catholicism, an instrument for the eventual conversion of Muscovy. At Jesuit instigation, Marina secured from her future consort the promise of the right to maintain Catholic priests at her court in Moscow, provision for the Catholic form of worship, and Catholic schools and churches in certain areas.

The Polish king, Sigismond III, was not taken in by the

Pretender's claims when he first met him at Cracow in 1604.
But he gave him a pension, recognized him as the legitimate
Tsar, and announced, although he did not officially declare war
on Muscovy, that Poles were free to engage themselves voluntarily in the small private army that the alleged Tsarevich was
forming. In the early autumn of 1604 this army of not more
than some 4,000 men set out for Moscow.

Godunov, in the meantime, unsuccessfully sought to discredit the Pretender and to whip up Orthodox sentiment against
'Latin and Lutheran heresy'. No distinction was made between
the two in Muscovy. But the demoralization of Godunov's
régime was evident in the Pretender's easy advance along a
general north-easterly route from Lvov to Moscow. His army
rapidly grew, swollen by warring bands of malcontents, homeless peasants, and, at a later stage, the Cossacks of the
Donetz.

Along the lower reaches of the Dnieper, the Donetz, the
Don, and the Volga rivers lived a nomadic mass of humanity,
sustained by plunder, brigandage, fishing, hunting, bee-keeping,
cattle-raising, and agriculture. The Cossacks were organized
in semi-military communities, in so far as they were organized
at all, and served as a Slav outpost against Crimean Tartars and
their Turkish overlords. But since the Cossacks were largely
formed of runaways from the serf régimes of Muscovy and
Poland and Lithuania, there was no clearly defined political
allegiance.

A few months after it had set out, the Pretender's army had
almost quadrupled in size. Godunov's forces, by contrast, disintegrated through mass desertion, boyar treachery, and defeatism. Who knew, perhaps they were indeed up in arms
against their legitimate ruler?

Suddenly, in April 1605, Godunov died – whether from a
stroke or from self-administered poison is another mystery of
the Time of Troubles. Two months later, after the virtual extinction or exile of the whole Godunov family, the Pretender
entered Moscow, to the uproarious tumult of the populace.
Maria Nagaya, the last wife of Ivan the Terrible and mother of
the true Dmitry, was brought from her convent and, doubtless

for reasons of prudence, acknowledged the Pretender as her long-lost son. He was crowned Tsar at the end of July 1605.

The new Tsar was assuredly one of the strangest rulers ever to occupy a throne. The French captain, Margaret, who was one of his commanders, found him aged about twenty-five, of small but vigorous stature, with an ugly, beardless face, a wart on his nose, and spiky red hair. His temperament was quite un-Russian in its liveliness. In the boyaral council his quick-wittedness marked him out. He took his responsibilities as Tsar lightly, doubled the pay of his officials and serving officers, was easily accessible to all with a grievance, confirmed and extended the privileges of the Orthodox clergy, and sought to protect the peasantry from such unfair conditions of re-engagement as would have been tantamount to slavery. There was an altogether fresh spirit of religious tolerance in the Kremlin and a welcome for foreigners. The Tsar even refused to use terror as a political weapon.

This period lasted less than a year. The Pretender's downfall was precipitated by the arrival of his fiancée, Marina Mniszek, from Poland. The future Tsarina's Polish entourage set Muscovite teeth on edge.

But this was not the only reason for the Pretender's downfall. Even if it had been his sincere desire, he could not have been of much use to the Catholic cause in Russia. To the Polish Jesuits and the Papacy, who were among his most ardent sponsors, this was necessarily a handicap. Far more important was the opposition of the Russian boyars. Their motive in using the Pretender was purely negative – to get rid of Godunov – and not to install another Tsar. Positively, their aim was a Tsar drawn from their own ranks, and this the false Dmitry obviously could not be.

The boyars' revolt, led by Vassily Shuisky, broke out in Moscow one May night in 1606. The Pretender was soon disposed of, his body burnt, and the ashes shot from a cannon to the four winds. Shuisky himself replaced the false Dmitry on the throne. He was not elected by the Zemsky Sobor, only nominated by his boyar associates and fellow conspirators. By an astonishing paradox, in the midst of lawlessness, Shuisky was unique among Tsars in that his autocratic power was explicitly limited by something of a constitutional document extracted from him by his fellow boyars. He pledged himself not to heed denunciation without first making due legal inquiry, not to carry out executions, not to punish innocent relatives of the guilty, and not to confiscate property without the approval of the boyaral council.

But these limitations were purely class-conditioned and had no relevance to any of the non-boyar population. They merely attempted to replace the autocracy by an oligarchic body of aristocrats. Outside that circle, protection of any kind was non-existent. The worst anarchy followed, as wider and wider classes of the population were drawn into the struggle. Their belief in the State as such, eroded by the grain-hoarding and speculation of the famine years, undermined by the turbulence of events in Moscow, was yet further shaken by the success of the false Dmitry in mobilizing the Cossacks and the disaffected elements of the south-western regions. Shuisky's seizure of the throne marked the end of one phase of the Time of Troubles – the internal boyaral struggle for power as against various forms of autocracy – and the opening of the second phase. Peasant revolt, combined with open foreign intervention by the Poles and Swedes, characterized the new period.

Shuisky's writ did not run much outside Moscow. Perhaps as much as a half of Muscovite territory was outside his control. Tula and Ryazan rose against him and defeated government troops from Moscow. Rebels besieged Nizhni Novgorod; Astrakhan declared its independence.

The emergence of a certain Ivan Bolotnikov emphasized the specifically peasant aspect of this new phase of turmoil. He was a former serf, whom a Tartar gang had swept away into slavery, where he had spent some years as helmsman on Turkish galleys. He then fled to Venice, and eventually regained Muscovy by way of Poland. As an agitator and revolutionary, he was the first to put into political terms the land-hunger and despair of the peasantry. His 'excellent charters' called on peasants and slaves to rise against their masters, to seize the estates of the *pomeshchiki*, and to overthrow the existing social order. For a time, especially in the south, this hardy call swelled Bolotnikov's following to a vast movement that carried him to the gates of Moscow.

There it began to disintegrate when, through the accession of the lesser provincial nobility, the movement lost its homogeneous peasant character. Bolotnikov's new recruits aimed only at unseating Shuisky – not at inaugurating a social revolution. Bolotnikov had to withdraw from Moscow to Tula. Then, although he had hitched his star to Tsarevich Peter, the soi-disant son of Tsar Theodore Ivanovich, the rebels eventually succumbed to Shuisky's siege in the autumn of 1607.

Hardly was the threat of a peasant revolution overcome when yet another centre of revolt arose. Under the leadership of a second false Dmitry – like the first, a man of unknown origin who gave out that he had not only escaped death at Uglitch in 1591 but also at Moscow in 1606 – a mixed array, composed in the main of Polish plunderers and swashbucklers, marched on Moscow from Starodub, some 300 miles to the south-west of the capital.

The second false Dmitry had less success than the first. Usually known as the 'Thief' or 'Scoundrel', because his claim was so patently spurious, his forces never quite reached Moscow. He established himself at Tushino, a village some six miles to the west, and there he set up a sort of 'anti-Government' with a separate court and appurtenances. Moscow and Tushino confronted each other. Many popular elements abandoned Shuisky for the Thief, for the latter was generous in his promises and bestowal of titles. This confrontation lasted about two years. Shuisky had lost the south, but the Thief was not strong enough to blockade the city and starve it into surrender. At the head of a force of more than 20,000 men, Jan Sapieha, the Polish noble and one of the powers behind the Thief, attempted to cut Moscow off from the north and the trading centre of the White Sea by seizing the strategic monastery of the Trinity of St Sergius. But its fortifications withstood the siege, to the enormous prestige of the Church as a national and centripetal force, and also helped to rally the pro-Shuisky forces in the north.

On the other hand, Shuisky came no nearer to overcoming

the Thief, and in this extremity he called in Swedish aid. In return for the renunciation of Muscovy's claim to Livonia, Charles IX of Sweden placed at Shuisky's disposal a force of 6,000 French, German, Scottish, and English auxiliaries under General De La Gardie. This in its turn provoked Sigismond III, King of Poland, to declare open war on Shuisky, lest the whole of the Baltic coast fall into enemy hands. In 1609 Sigismond laid siege to Smolensk, the classical key to Western Russia.

This precipitated the final phase of the Troubles. The Thief's camp disintegrated, Tsar Shuisky was deposed, and a Polish garrison was master of Moscow. Meanwhile, negotiations opened between the boyaral council and Sigismond on the terms that would make a Polish Tsar acceptable to Moscow. In the upshot, the council recognized Sigismond himself as Tsar. This was perhaps the nadir of Muscovite misfortune. Those Russian nobles opposed to the Poles were at loggerheads with the Cossacks, ostensibly their allies. Smolensk fell to Sigismond; and the Swedes had seized Novgorod. Apparently no force existed able to arrest the utter disintegration of the State. The only comparison possible is with the territory that the Bolsheviks controlled at the end of the First World War, when they too were simultaneously subjected to foreign intervention and civil war.

The solution eventually came from a combination of the influence of the Orthodox Church and that of the towns, primarily those to the east of Moscow. The first nationalist levy to attack and besiege the Poles in Moscow in 1611 was formed largely of the provincial nobility. But it also included a large force of Cossacks, and the Poles were able to instigate dissension between the two elements to such good effect that the levy dispersed.

The second attempt, made in the following year, was rather better prepared. It took its inspiration from the leaders of the most famous and most ancient of all Russian monasteries – that of the Trinity of St Sergius, already prominent through its sixteen months' resistance to the besieging troops of the Polish princes, Lizovski and Sapieha. The cellarer and bursar of the monastery, Abraham Palitsyn, has left a remarkable account of this period. Under the leadership of the Church, a stream of appeals was directed to many of the towns of the Russian east – Kazan, Perm, Nizhni Novgorod – exhorting them to rise against the invader and the usurper. In Nizhni-Novgorod the response, organized by the mayor of the town, a merchant named Kosma Minin, was particularly strong. A second national levy, made up of townspeople under Minin and of the lesser nobility under Prince Pozharski, was rapidly formed. Although there was a last-minute dissension with the Cossacks, still attacking the Polish garrison in Moscow, the two forces were able to combine in a final supreme effort that freed the city.

Early in 1613 the dynastic problem was solved by the summoning of a Zemsky Sobor of the most representative type. Its task was to elect a Tsar, and this, after much deliberation, it succeeded in doing. The choice fell on Michael Romanov, a member of the celebrated boyaral family, which had not, however, compromised itself irretrievably during the Troubles. It was satisfactory both to the Sobor and to the Cossacks. The election was not altogether a break with the past, for Ivan the Terrible's first wife, Anastasya, had been connected with the Romanov family. Still less did the mere fact of election imply any temptation to follow the Polish system of an elective monarchy. Michael Feodorovich Romanov was crowned Tsar with all his autocratic prerogatives intact.

CHAPTER 6

Muscovy in the Seventeenth Century

THE election of the first of the Romanov dynasty by no means marked the end of the Troubles. The Poles and Swedes still occupied parts of Muscovy. The treasury was empty. Large areas of the country fell prey to wandering bands, and the disruption and flight of the population resulted in a steep decline in agricultural productivity. On the estates of the monastery of the Trinity of St Sergius, for example, arable land, which had formed 37·3 per cent at the end of the sixteenth century, fell to 1·8 per cent by 1614–16. Over the years 1621–40 it rose again, but only to 22·7 per cent. Not until the 1640s was the upturn evident.[1] The poverty of the court and the government can be seen in the Tsar's plea, a bare few months after his election, to the Stroganov family. In humiliating terms he had to ask for a loan in cash and kind – fish, salt, grain – to meet the urgent demands of his officials and military personnel. In these inauspicious circumstances the Romanov dynasty entered on its long career.

But the country was already showing remarkable resilience in surviving the virtual collapse of all authority. In the three-quarters of a century that followed the Time of Troubles, Muscovy was more prominent than ever on the European map. The Swedes, during the Thirty Years War, were the first to maintain a permanent embassy in Moscow. Danish and Dutch envoys followed, as did representatives of the Great Elector of Brandenburg. Plenipotentiaries came from England, Poland, and the Holy Roman Empire as the occasion demanded. The reverse movement also developed, as did trading relations with England and Holland. Not only was Archangel used, but also the Swedish Baltic ports of Riga, Revel, and Narva. Along with this development went the growth of a sizeable foreign quarter

1. Peter Lyashchenko, *History of the National Economy of Russia*, English translation, New York, 1949, p. 201.

in Moscow. This foreign quarter was ultimately to bring before Muscovy the perplexity of its historical destiny, fated to prove a fruitful source of intellectual tension.

The pre-condition of all this recovery was a united country. This was the chief and most important result of the Time of Troubles. The peasant question, of course, was by no means solved. In fact, the intensification of serfdom produced a truly revolutionary movement in Stenka Razin's uprising from 1667 to 1671. New sources of unrest were the urban poor and the conquered populations of Siberia and the east. But none of this meant that the dominant class of *dvoryane* (as the old *pomeshchiki* came to be known) and merchants was not fundamentally united and in control of the State. Here, too, there were conflicts and divergences, but barely, if at all, did they involve intervention by the mass.

The Romanovs reaped what the sons of Rurik had sown. Decimated, evicted, and uprooted by the onslaughts of Ivan the Terrible, Boris Godunov, and the Troubles, the boyars no longer represented a political or economic force of any worthwhile magnitude so long as the central power remained strong. Their advantage of birth might still bring with it social preferment and a seat on the boyaral council, but these were the limits of their prerogatives in normal times. The end of the old order was nowhere more graphically shown than in the abolition of the system of *mestnichestvo* – the system whereby inherited family rank and status determined the appointment of army commanders, ambassadors, and various court functionaries. *Mestnichestvo* was already falling into disuse in the first half of the seventeenth century, but when the books of rank recording the genealogical tables were solemnly burnt in 1682, the break with a characteristic tradition of old Muscovy was complete. What remained was a strong clan feeling.

The Zemsky Sobor was in more or less permanent session from 1613 to 1622, ruling in cooperation with Tsar Michael and his father, the Patriarch Philaret. The latter was released from a Polish prison in 1619. The Sobor had a voice in deciding matters of war and peace, in authorizing new fiscal measures, increases in taxation, and generally in helping to consolidate and execute the policies of the central authority. By 1622 its main task was done. Thereafter it was only convoked at lengthy intervals – in 1632, 1637, 1642, and 1649, for example, and then with a much diminished role. It had no further purpose to fulfil.

The new Muscovy of the seventeenth century was already a State organized in the interests of the *dvoryanin*. 'Organized' is indeed the operative word, to which the seventeenth century was to give an ever more precise meaning. There was organization of the population in the interests of State service; organization of trade and industry; organization of government; and every new economic development, such as mining or manufacture, automatically fell under the aegis of the State. Indeed, there was a constant effort to bring all the activities of the population under this aegis, for at the back of this policy stood the needs of war, engaging all the resources of the State.

The handling of the supremely important peasant question shows this at its clearest. The practice had already grown up of designating certain periods during which runaway peasants could be sought out and forcibly returned to the estates they had quitted.[1] By 1642 the set interval was ten years. But this was far from satisfying the smaller landowners, whose estates, in their absence on service, were frequently robbed by the larger landowners in search of manpower.

A more lasting though still unsatisfactory solution had been in the making since 1619. A census was then ordered, with the aim of bringing back the runaways and also of unmasking those free peasants who, during the Time of Troubles, had 'commended' themselves to some large landowners, lost their freedom, and thereby relieved themselves of the liability to taxation. The work went at a snail's pace and then came to a complete standstill: all the results were burnt during the great fire of Moscow in 1626. But it was resumed, and quickly concluded in 1628.

This was the basis of the famous code of 1649 which entirely abolished 'fixed years' in the search for runaways and established the landowner's legal, unlimited right to them. The prin-

1. See p. 49.

ciple was that the peasants should be returned to the estates where they had been registered in the census returns. There was also provision for fining those landowners who retained peasants to whom they were not entitled. In the circumstances of the time, much of this decree remained a paper right. In 1658 it had to be reinforced by a further decree that introduced special officials to seek out runaway peasants; and in 1661 and later years, flogging and a severe scale of monetary fines were prescribed as punishments for those landowners who took in runaways. Also, as a penalty for such infringement of the law, the guilty landowner had to surrender four of his own peasants for each runaway peasant whom he illegally harboured.

The code of 1649 was the climax to a development extending over the previous century and a half; and during the next two centuries it remained substantially unaltered as the predominant factor in the Russian economy – not only in agriculture, but also in the State industries of the seventeenth, eighteenth, and nineteenth centuries. Its effects and ramifications were so far-reaching, and so intimately identified was the State with serf-dom, that reform or abolition of the institution as such was the starting-point for almost every liberal, radical, or critical observer of Russan conditions.

In the middle of the seventeenth century it served the dual purpose of providing the small and middling landowners with a ready supply of attached labourers, attached not only to a particular estate but also to the estate-owner. The eighteenth century saw probably the apogee of the landowner's powers over his serfs. But in the seventeenth century, despite certain anomalies and contradictions in the law, a landowner could already sell his serfs, exchange them, punish them, split up families, and, in general, treat the serfs as chattels. He could not, however, turn them into slaves, for this would deprive them of their taxable status. It is true that the code of 1649 expressly warned the landowner not to kill, wound, or starve his serfs, but there was very little sanction to prevent such maltreatment.

In the main, the landowner used the serf as a performer of *barshchina* – labour – on his estate. This might take up between one-third and one-half of the serf's working time. The system operated predominantly in the central regions of Muscovy. The north-east, Siberia, and the steppe-lands of the south were still largely untouched. Elsewhere the distinction between the slave and the serf grew gradually less and less.

Neither the serf nor the poor townsman ever reconciled himself to the next exactions of the State and the landowner. The seventeenth century was marked by a number of elemental movements of revolt, as well as many sporadic individual outbursts. A tax on salt led to an upheaval in Moscow in 1648; in 1650, revolt came to Pskov and Novgorod; in 1656, the Tsar's debasement of the currency (he replaced silver coins by copper) provoked further unrest throughout all Muscovy and led to the deportation and torture *en masse* of many thousands of the disaffected.

The most significant of these outbreaks was the movement led by Stenka Razin – a Don Cossack. In the vast area of east and south-east Muscovy he incited a peasant revolt, aimed not so much against the Tsar this time but against the boyars and landowners. It was the successor to Bolotnikov's movement of the Time of Troubles and the herald of Buturlin's revolt under Peter the Great, and Pugachov's even more tumultuous upheaval under Catherine. Razin's first raids set aflame the lower Volga region and the shores of the Caspian Sea. Then, in 1670, he turned northwards and conducted a veritable *jacquerie* against the land-owning classes, winning over thousands of serfs and also many of the tribes not yet reconciled to Muscovite exploitation – Bashkirs, Kalmyks, Mordvinians. The movement spread through the whole Volga region and as far northwards as the Oka. But regular Muscovite troops defeated Razin near Simbirsk on the Volga, and in 1671 he was executed.

The transformation of the still nominally free peasant into an attached serf was only one aspect of the increasing organization and even militarization of Muscovite society – a tendency already apparent in the sixteenth century. Now it received fresh impetus from the State's need for maximum revenue in addition to the perennial requirements of border defence.

From Ivan the Terrible and ultimately from Ivan the Great the Romanovs inherited a system of government characterized by thirty to fifty chancelleries, or Prikazi, which functioned through officials both in the capital and in the provinces. The jurisdiction of the Prikazi was ill-defined. Some had a territorial and others an overlapping functional basis. But there was in general a tendency towards concentration and centralization. The same development was manifest in the administration of the provinces, where Ivan the Terrible's encouragement of local self-government succumbed to the growing powers of the *voevoda*, the Tsar's representative. He more or less embodied the resurrected spirit of the old *kormlenshchik* of the sixteenth century, with all his abuses and exactions.

In their reform of the army the Romanovs were more venturesome. Its inefficiency was too glaring to be left unremedied. The system of mobilizing the *dvoryane* as cavalrymen in military emergencies was retained. The real novelty lay in creating a standing army, recruited from and officered by foreign mercenaries, many of whom were veterans of the Thirty Years War – Germans, Swedes, Poles, Scots. There was also expansion of *streltsy* regiments, made up of men quartered in special parts of Moscow and the other chief towns who would in normal times pursue an occupation in handicraft or trade. But they could be rapidly mobilized in time of war. The *streltsy* received a fixed annual salary, and through their closeness to the throne and their caste-like character were at times politically important – especially if the salary was late.

Side by side with military reform went the division of the population into a number of theoretically fixed and hereditary classes or estates. Each estate entailed definite obligations and carried with it a certain status. Military service had always been the *raison d'être* of the *dvoryanin*. Now this became a hereditary burden. There was no escape. As compensation, however, the *dvoryanin* could henceforth join the ranks of the new nobility. Furthermore, a law of 1642 and the code of 1649 permitted only the *dvoryanin* to own land worked by serfs. In response to *dvoryanin* pressure also, the Church's land-owning was clipped. There was no question of secularizing ecclesiastical estates at this period, but limits were set to their future expansion. Even so, according to Samuel Collins who was in Moscow in the early 1660s as physician to Tsar Alexis, the Church still remained owner of almost two-thirds of Muscovy. According to Kotoschichin, one of the earliest of Russian social critics, the Patriarch owned 7,000 peasant homesteads, the bishops 28,000, and the monasteries 83,000.[1]

The service nobility had every reason to be satisfied with the concessions made to it in the code. Hardly less favoured were the merchants, townspeople, and tradespeople in that they received a monopoly of such industrial and commercial activities as existed. But they certainly did not enjoy commercial freedom. Burghers were forbidden, on pain of capital punishment, to move from one town to another or to sell shops and houses to members of another class. The government's aim, as in the case of the *dvoryane*, was to retain all potential sources of taxable revenue and to strengthen existing sources. The landed men of service, with some exceptions, were forbidden to engage in trade or industry. Both merchants and landowners were essentially servants of the State.

The seventeenth-century Muscovite town was practically indistinguishable from the surrounding countryside. There was nothing comparable to the Western-European town as a middle-class bastion; still less was there a middle class or an artisan class grouped in guilds. The town and the surrounding countryside used exactly the same methods of handicraft production. The peasants were well able to satisfy from their own resources their requirements in clothing, housing, and household utensils. In many cases, especially in the south and south-east, the towns were not much more than fortified strongpoints, largely inhabited by the military.

For all that, the towns were a centre of handicraftsmen who formed a significant portion of the urban population. In Moscow in 1638 there were 2,367 artisans, in Kolomna 159 (twenty-two per cent of the adult population), in Mozhaisk 224 (forty per cent), in Serpukhov 331, in Kazan 318 (more than fifty per cent), in Novgorod about 2,000, in Nizhni Novgorod about

1. For Kotoschichin, see p. 89.

500. The non-industrial urban groups included such people as coachmen, porters, apothecaries, midwives, bloodletters.

It is possible to distinguish various categories of urban workers. Most numerous were the producers of foodstuffs – bakers, brewers, distillers, fishmongers. A second large group was comprised of clothing workers. These men were hatters, furriers, coatmakers, tanners, tailors. They worked mainly with the wool and flax processed by rural spinners and weavers. Lastly came the metal workers, who produced both articles of military use – swords, armour, axes, arrows – and household utensils. These men were tinsmiths, locksmiths, and needle-makers. It seems that the level of technique was extremely low. According to one observer, not even saws were used by the carpenters. The boards and planks were hewn straight from the log with an axe. Generally speaking, the urban artisans and craftsmen were the town poor. Even such a skilled worker as the silversmith sold his products for not much more than the cost of the metal. Labour was very poorly rewarded.

Precisely because there was so little differentiation between town and country, all the town's activity was duplicated on the larger estates and monasteries, where, in fact, there was probably a higher degree of specialization as well as a greater diversification of crafts. The work might be performed by serf artisans; but there were also 'free' artisans who received cash for their products, either per unit of product or per half-year.

Much of this work was for immediate consumption and use. But some of it was for the market and the requirements of a money economy, on the pattern manifested in the sixteenth century. Local and regional fairs in hamlets and villages were often arranged by monasteries to coincide with Church holidays. An enormous variety of goods was bought and sold – farming implements, animals, vegetables, fruit, poultry, clothing, salt, and meat.

The most important trading centre was undoubtedly Moscow. It stood at the centre of the new trade-routes criss-crossing the country: the road to Western Europe by way of Smolensk, Vitebsk, Riga; the immensely important route to Archangel; while the Moskva and Oka river system joined Moscow with the Volga as far south as Astrakhan, and also with Siberia. There was also a radial network of shorter routes linking Moscow to such towns as Tula, Ryazan, Kostroma, Vladimir, and Kaluga. Almost one-third of the twenty-per-cent tax levied in 1634 on the whole commercial and industrial part of the population came from Moscow alone. This further illustrates the capital's commercial predominance.

Here were transported the products of fishing, trapping, and hunting, the agricultural wealth of the central and southern districts, iron-ware from the Ural and Tula ironworks. Here were two bazaars every week and ever-open market-places. Foreigners in Moscow were of one voice in their surprise at the number of shops in the city. This was noted by the Jesuit father, Possevinus, at the time of the false Dmitry, and by the Dutchman, Kilburger, towards the end of the seventeenth century. Moscow, said the latter, had as many shops as Amsterdam. But the comparison is highly misleading. Many of the so-called shops were not much more than booths, stands, trestles, or barrels with a few articles on top. One shop in Venice had more goods than a whole row of shops in Moscow, wrote Possevinus.

Commerce was concentrated around the Kremlin and in the specialized trading quarter – the *Kitai-gorod*. The 'shops' were grouped in rows in accordance with the nature of their stock. Pedlars circulated between the rows, hawking trinkets, miscellaneous wares, and foodstuffs. The retailers were not all Russian; they might also be from Western Europe or the Orient. But Muscovite traders sternly contested this privilege and finally secured its abolition in 1667.

Moscow was the centre of foreign as well as of internal trade. Silk from Persia and China, wine from France – Muscovites favoured Anjou and Bordeaux – were some of the more noteworthy items. Altogether, the seventeenth century saw considerable expansion in Russian imports and exports. This is well illustrated in a curious by-product of the period – the development of a postal service with the West. It was first handed over to a Dutch entrepreneur who organized two main postal routes – Moscow-Smolensk-Vilna and Moscow-Novgorod-Pskov-

Riga. From Vilna the post went on farther to Königsberg, Berlin, and Hamburg. Letters took four weeks from Moscow to Hamburg; to Danzig a week less. The letters arriving in Moscow were first opened at the rudimentary foreign ministry so that no private individual should know what was happening inside or outside the country before the court. This was one of the most frequent criticisms of the postal service. It was also accused of irregularity, and its officials of diverting to their own pockets any valuables or currency found in the opened packets. But it undeniably fulfilled a service.

In the continued absence in the seventeenth century of a port in the Baltic, Archangel remained the chief opening to the West. About the middle of July, the merchants of Muscovy set out on the fourteen-day trip northwards. So great was the efflux that activity in Moscow slackened perceptibly. In July also, the foreign ships arrived for the annual fair. This lasted until September. Despite the long and hazardous sea journey, it was cheaper than the overland route, which would have involved excessive tolls and transit taxes.

The Russians exported grain, leather, furs, canvas, hemp, flax, lard, wax, and caviar. For much of the century their favoured imports were such materials of war as iron, powder, and arms from England, Sweden, and Holland. Special purchasing agents were appointed to procure the muskets, daggers, shot, and books on military engineering and fortifications that the armies of Muscovy required. Later, articles of luxury and adornment came to occupy a large part of the trade. In 1671, imports through Archangel included diamonds, articles wrought in gold and silver, gold coins, 28,000 reams of paper, 10,000 German hats, 837,000 pins and needles, wines, spices, muscatel, and herrings.

There was as yet no merchant marine to speak of, so Muscovy was entirely dependent on foreigners for its trade with the West. But this source of weakness was partly cancelled out by the competition for what was a very profitable two-way traffic. And it was this that in 1649 enabled the second Romanov, Tsar Alexis, to abolish the privileges enjoyed by the Russia Company. It had its humorous side, for the chosen pretext – in Muscovy, of all places – was the execution of Charles I. Alexis would also not recognize Cromwell. When the Stuart dynasty was restored, the English plausibly argued that they had thereby atoned for the sin of regicide. To no avail – Muscovy no longer needed the English as it had a century earlier.

Owing to the multiplicity of retail outlets and of producers, the poor communications and the sheer distances involved, home trade could not be regimented to the same extent as foreign. But they were both dominated by three corporations of merchant-entrepreneurs, the chief of which were known as 'guests'. They worked in intimate cooperation with the Tsar as the foremost representatives of mercantile capital. In all Moscow there were about thirty 'guests'. Kilburger describes them as 'royal commercial counsellors and factors' – 'a greedy and pernicious fraternity', who would be the first to get their necks broken if ever riots broke out. They enjoyed such privileges as tax immunity, the right to unrestricted travel abroad, and the right to distil vodka for their own consumption. Basically, they functioned as the administrators of the customs, the State fisheries, and salt works. They purchased goods for the Tsar and also traded on their own account; they operated leases and also handled the Tsar's foreign and internal trade. One 'guest', a certain Nikitinov, was prominent in textiles, salt, and fish. He also owned a fleet of vessels that plied up and down the Volga as far as Astrakhan. Others were Gruditsyn and Voronin, who owned villages, stores, ships, ironworks, and functioned as army contractors. The most noteworthy 'guest' was the Tsar himself, who could buy imported goods at the prices asked and then, by way of the corporation, dispose of them at inflated prices on the home market. Conversely, through the operation of the royal monopoly he could buy advantageously on the home market and secure the market price at Archangel. In this the Tsars of the seventeenth century were following the tradition set by Ivan the Terrible and Boris Godunov.

Trade, as a source of wealth, was paralleled by the exploitation of such natural products as potash. Here, the boyar Morozov was the pioneer in large-scale enterprise. He used the labour of thousands of serfs to work the deposits. Mining and metallurgical industries, directed by foreign masters and specialists, were another significant development. On a larger scale, this continued the policy initiated by Ivan III and Ivan the Terrible, and anticipated the efforts of Peter the Great – with the same aim of lessening dependence on the foreigner, combined, at first, with a pronounced military emphasis. Instead of importing manufactured goods and exporting raw materials, Muscovy, it was hoped, would produce her own manufactured goods.

In pursuance of the Romanov policy of developing the army, the first foreign-managed enterprise of importance was an iron-works at Tula. Here, a Dutchman, Andreas Vinnius, established in 1632 on State land a plant that would supply the Government with cannon, cannon-balls, and musket barrels at agreed prices. In 1644, Vinnius was joined by a second Dutchman, Peter Marselis, who was granted permission to establish iron-works in certain northern districts. There was also a third iron-works near Moscow that belonged to the Tsar. This, too, was intended for the casting of cannon and other army require-ments. The arrangement was in the nature of a concession for a limited number of years, after which the establishment re-verted to the State. The foreigner supplied the technical skill and undertook to instruct Muscovite apprentices and the State supplied the labour, consisting of the serfs directly employed in the enterprise, or those obliged to deliver fuel or perform other auxiliary services. This was the origin of the 'factory peasants' of the eighteenth century.

On this general model other enterprises developed during the seventeenth century – a glass works, leather and silk factories, a paper mill. Not all of these were successful and in any case they were all on a very small scale. But they did at least signify the establishment of a type of industry fundamentally different from manorial handicrafts.

WAR IN THE WEST

Muscovite territorial losses during the Time of Troubles made
the country more of a Eurasian state than ever before. The
position in Siberia and east of the Urals remained unaffected,
whereas important towns in the west, such as Novgorod and
Smolensk, were still held by the Swedes and Poles respectively.
Muscovy's seventeenth-century wars were largely devoted to
restoring some sort of balance (although, of course, penetration eastwards never ceased). In the unremitting impulse westwards, the balance of power slowly moved in favour of Muscovy and against Poland. In fact, events on the western borders
of Muscovy seem to foreshadow the Polish partitions of the
late eighteenth century.

The immediate task after the Time of Troubles was to clear
Russian soil of the Swedes and Poles. The former were the
easier to deal with. After desultory fighting, which included
a vain Swedish assault on Pskov, the Swedish occupation of
Novgorod and its surrounding area was ended by the Peace of
Stolbovo in 1617. Dutch and British mediation, for commercial motives, and a British loan helped to ease the settlement.
Muscovy paid an indemnity of 20,000 silver roubles, in return
for which Sweden evacuated Novgorod and renounced its
claim to the Muscovite throne, while retaining the southern
coast of the Gulf of Finland. This established Swedish contact
with its Estonian possessions and left Muscovy still cut off from
the Baltic. There was a resumption of war against Sweden in
1656 over the control of Poland and Lithuania, but in 1661 the
Muscovites were beaten back.

Poland proved an even harder nut to crack. With Smolensk
still in their possession at the end of the Time of Troubles, the
Poles were able to launch a new offensive that brought them
close to Moscow in 1618. But the capital withstood the assault,
and this did in fact mark the last occasion when Polish troops
entered central Muscovy. At the end of the year an armistice
for fourteen and a half years was concluded at Duelino. Poland
retained the provinces of Smolensk and Seversk, but did not renounce a claim to the throne of Muscovy, a claim that stemmed

originally from the Troubles. Happily for Muscovy, the expiry of the armistice in 1632 coincided with an interregnum in Poland when the Polish king, Sigismond Augustus III, died. Muscovy tried to exploit an apparently favourable opportunity to retake Smolensk, but the siege was a fiasco. In 1634 an 'eternal peace' was concluded, as distinct from the earlier armistice. Apart from the final abandonment of the Polish claim to the Kremlin, it did not change the *status quo*.

The 'eternal peace' lived barely longer than the armistice. It came to grief in the 1650s in the Ukraine (or Little Russia), where a most complex conflict was building up between the Poles, the Russians, and the Cossacks. With fine impartiality, the Cossacks would fight for the Holy Roman Emperor against the Turks, for Muscovy against Poland, for Poland against Muscovy. Their aim was to live unhampered by any ties save those of military brotherhood. However, as Muscovy gradually expanded southwards, and as Poland, after union with Lithuania in 1569, also began to bring the Lithuanian territories in the Ukraine under some form of regularized control, Cossack autonomy came under a two-pronged attack. The conflict was further envenomed by the formation of a Uniate Church, under Jesuit inspiration. The Uniates accepted the authority of the Pope but retained the Orthodox form and ritual. The conflict of Orthodoxy and Catholicism thus moved south.

In 1651, after more than a decade of inconclusive Cossack wars and uprisings against the Poles, Bogdan Khmelnitzki, the *hetman*, who had also distinguished himself by his merciless attacks on the Jews of the Ukraine, had to appeal to Tsar Alexis for help. Would he make the Ukraine autonomous under Muscovite suzerainty? Muscovy, mindful of its uniformly unsuccessful wars against Poland, hesitated, but finally agreed. In January 1654 the Cossacks swore their oath of allegiance to the Tsar. At Pereyaslavl a treaty was solemnly and emotionally concluded between Khmelnitzki and Buturin, the envoy of Alexis. The Ukraine thereby received confirmation of its autonomy, and the right to entertain independent diplomatic relations, except with Poland and Turkey.

This treaty by no means settled the future of the Ukraine,

but this aspect of the Russo-Polish conflict was quickly over-shadowed by the outbreak of war between the two countries. For the first time, Muscovite superiority was evident. Smolensk quickly fell, to be followed in 1655 by Vilna, Kovno, Grodno, and large parts of Lithuania. The Muscovites had not only Cossack help, but Swedish also, for Charles X, invading from the north, captured Warsaw and Cracow and proclaimed himself King of Poland. This was too much of a good thing for Alexis, who himself aspired to the Polish crown. He broke off his Polish campaign and turned on his ally, hoping to secure part of the Baltic coast. But his fight against Sweden brought him no success. Muscovy came to grief at Riga. Finally, confused by the conflicting claims of the Ukraine and the Baltic, Muscovy concluded with Sweden the Peace of Cardis in 1661. All Livonia was returned to Sweden and the latter's possession of Estonia was confirmed.

In the south, Khmelnitzki died in 1667, to be commemorated by a grandiose equestrian statue in Kiev, capital of the Ukraine. Under his successors, there were rapid and bewildering changes in Ukrainian foreign policy as Moscow systematically disregarded its promises of non-intervention and began to move its *voevodi* and their garrisons south-westwards. At close quarters, Muscovy did not seem such an improvement on the Poles.

The consequence was a renewal of the Muscovite-Polish struggle. Some nine years' fighting ended in 1667 with the armistice of Andrussovo, which split the Ukraine. The left bank, the area east of the Dnieper, fell to Muscovy, together with Smolensk and the province of Seversk. Kiev too returned to Muscovy – after an absence of some four centuries. Kiev stood on the right, western bank of the river and was to have been relinquished by Muscovy after two years. But this provision of the armistice was never carried out. In the following year, John Casimir, the Polish king, renounced his throne, and withdrew to the more congenial soil of France and to the lands bestowed on him by Louis XIV.

Muscovy's possession of the left-bank Ukraine and of Kiev could be uncertain only so long as it depended on the armistice of Andrussovo, and even more, on its violation. Their final acquisition was bound up with the formation under papal patronage of the anti-Turkish coalition of the 1680s, consisting of Poland, the Holy Roman Empire, and Venice. Muscovy's two campaigns in the Crimea (in 1687 and 1689) were scarcely successful, but the essential preliminary *rapprochement* between Poland and Muscovy involved the final cession of the conditional gains made under Andrussovo. Just over a century later, in 1793, the second partition of Poland united the right bank of the Dnieper with the left. Not only in a territorial sense did the 1680s significantly foreshadow the future. It was also the first time that Muscovy took part in a European coalition, an ally of some consequence.

Expansion to the east, which even the Time of Troubles did not interrupt, encountered fewer difficulties than expansion to the west. During the seventeenth century, Muscovite bands of military adventurers, many of whom were criminals on the run, conquered all eastern Siberia and the whole of the Lena valley. There followed the subjugation of the extreme northeast, beyond the Kamchatka peninsula and the Sea of Okhotsk. By the end of the seventeenth century, Muscovy, landlocked to the west, was already bordering on the Pacific. Penetration beyond the Amur River was limited by the Treaty of Nerchinsk with China in 1689. More regular authorities from Muscovy followed the adventurers. Strongpoints and fortified centres were established, as well as customs-houses to levy and collect tribute from the primitive tribes of the region.

WESTERN INFLUENCES

In the wake of foreign policy and foreign technicians came a slow seepage into Muscovy of Western ideas. Greek Orthodoxy had hitherto dominated such little intellectual life as existed. The Renaissance, Reformation, and Counter-Reformation passed Muscovy by completely. Apart from certain obscure Judaizing heresies, involving the denial of the Trinity and of the divinity of Christ, the observance of Saturday rather than Sunday as the Sabbath, the rejection of the Madonna as an object of worship – heresies that seem to have emanated from

Novgorod at the end of the fifteenth century and then infiltrated into Moscow – the Orthodox Church was able to enforce a more or less complete separation from developments in the West. Whereas to the West, Greece signified Hellenic antiquity, to Muscovy it signified Byzantium. Any incipient intellectual penetration was sternly repressed. This was carried to such lengths that Tsar Michael, in the first half of the seventeenth century, would have a golden jug and basin beside his throne in the audience room of the Kremlin. After receiving a Western ambassador, he would then wash away the stain of occidental pollution.

It was noted by Masaryk in the nineteenth century that European ideas necessarily had a revolutionary effect in Russia. The same was true of the seventeenth century. This first became apparent in the field of religion and followed on the conquest of the Ukraine. Politically, it was desirable to bring Muscovite and Ukrainian Orthodoxy into the closest possible harmony, in order to facilitate the absorption of the conquered territories. Moreover, the Ukrainians, who came directly under the authority of the Constantinople Patriarchate, had remained closer to the original Byzantine usages. On the whole, their clergy were on a higher level than the Muscovites'.

There was thus considerable inducement to reform Muscovite Orthodox liturgy and ritual and to accept the offer of the Ukrainian clergy to help in this task. Did one cross oneself with two fingers or three? Did one recite a double or a triple Hallelujah? Such trivia as these provoked the most intense conflict. Nikon, the masterful and imperious Muscovite Patriarch, began to introduce the reforms in 1653, and at once met with fierce opposition. This was not only religious, of course. The Orthodox Church, since its earliest days, had identified itself very intimately with the life of the people and State. The 'white' parish clergy lived the life of their neighbours in the villages. The Church was undoubtedly at a low intellectual level, but – perhaps for that very reason – at one with its congregation, so that Nikon was in fact launching an offensive against the long-cherished popular associations of Moscow – The Third Rome, the heir to Byzantium, the Christian bulwark against the Mongols – and against all the self-sufficiency of Muscovy. Again, to accept ideas from the West, as distinct from techniques, savoured somehow of treachery; for the West, from the days of the Teutonic Knights onwards, had always confronted Muscovy as an aggressor. This was the reality that made sense of the fanatical opposition that faced Nikon. Hitherto, religious uniformity had preserved Muscovy from the type of religious persecution practised in the West in the sixteenth century. Now this happiness was lost as the reforms were pressed home with fire and sword. Dissenting ecclesiastics were exiled, if not burnt at the stake. This latter was the fate of the arch-priest Avvakum, the most outspoken opponent of the reforms. The climax came with the attack on Solovetzky monastery, which had to endure a siege of eight years. This was a foretaste of the sort of opposition that Peter the Great's Western-inspired reforms would provoke half a century later.

Thus Muscovite isolation, to some limited extent, was already breaking down in the seventeenth century. It is significant that this period produced Prince Khvorostinin, the first Muscovite free-thinker of Catholicizing tendencies, sometimes considered a remote precursor of Tchaadaev in the nineteenth century. It is also no coincidence that the seventeenth century provoked a minor Muscovite diplomatist, Gregory Kotoschichin, to pen a highly unflattering description of the institutions and customs of his native country: illiterate and inarticulate boyars resting their long beards on the table of the boyaral council chamber; criminality in Moscow; loveless marriages and dissolute family life; universal arrogance, corruption, and oppression.

This indictment was actually written in Stockholm, whither Kotoschichin had fled. But there were also sources of comparison in Moscow itself. Of these, the most important was the Nemetzkaya Sloboda, that is, the German Settlement. (To the Muscovite masses all foreigners were Germans; in actual fact many of the 'Germans' were Scots and Dutch.) The first foreign settlement grew up in the vicinity of Moscow in the reign of Ivan the Terrible. It was destroyed during the Time of Troubles and abandoned. During the next three decades or so it underwent a renaissance inside the capital itself, where foreigners could purchase houses and even establish their own churches. But in 1643, following a fracas at one of the churches, the Patriarch ordered its demolition. The eventual result was the creation of a new Nemetzkaya Sloboda outside the capital, in which were assembled all the foreigners evicted from their previous residences. It quickly developed into a handsome suburb with three Lutheran churches and one Calvinist – Catholics might only worship privately – and a population of several thousand.

This enclave proved to be a centre of Western influence and Western amenities. The impact did not go very deep and it certainly did not go beyond a small stratum of the Muscovite aristocracy. For all that, the Sloboda is not unimportant historically. To the West as a source of power was added the concept of the West as a source of comfort and amenity.

Through their influential position in Muscovite society, the 'Germans', who were military men, merchants, teachers, artisans, doctors, apothecaries, paper manufacturers, iron-smelters, glass-workers, formed something of a European community, asserting a system of values that transcended mere practical usefulness. Western-style music would be organized to accompany the banquets of the Kremlin; the residences of leading boyars would house European furniture, clocks, mirrors, pictures. The boyars and the Tsar would ride in carriages upholstered in velvet and fitted out with glass windows. One of the most remarkable innovations of the period was the erection of a building in the village of Preobrazhenskoe for use as a theatre. The director was Johann Gregory, the pastor of a German Lutheran church in the Sloboda. The Tsar's religious scruples were overcome by reference to the example of the Byzantine emperors. The first performance of a stage play in Muscovy took place in October 1672. It was a Biblical tragicomedy on the theme of the Esther story, composed by Gregory, and performed by pupils of the German school, with specially trained Muscovite serf-musicians providing the musical accompaniment. It lasted ten hours. The next performance was of the tragedy *Judith and Holofernes*. With the establishment of a 'Palace of Amusements' within its walls, the

Kremlin itself soon witnessed theatrical performances. The conduct of the theatre was entrusted to the boyar Matveyev, a true connoisseur of European recreations. There followed the additional novelty of representations in Russian – in the case of the German pieces, Alexis had had to rely on an interpreter for a running commentary on the action – and the foundation of a theatre school, under the direction of Pastor Gregory. The students were all male. Not until a century later did women take the female parts. As an index of Russian cultural immaturity, this general situation may be usefully contrasted with the Restoration comedy of contemporary England or the French theatrical world of Racine and Molière.

Alexis's own attitude to these Western-European innovations fluctuated. A definite reaction followed his death in 1676. The activities of Pastor Gregory, Matveyev, and the Palace of Amusements were all suspended. Even so, intensified contact with the West was moulding a generation of Muscovite boyars into something very different from Kotoschichin's portraits. Thus, although the circulation of Latin and Polish works and the importation of foreign literature were formally forbidden in 1672 – Moscow already had its own printing-press, the Government argued – a man such as Matveyev had a library in some four or five European languages, and maintained with his wife, a Scottish lady named Hamilton, something of a European *salon*. Of comparable significance was Vassili Golitsin, whose house, said a French diplomatic envoy, La Neuville, was one of the most magnificent in all Europe, recalling the court of an Italian Prince. Another such boyar was Ordin-Nashchokin, the Muscovite equivalent of Foreign Minister.

The same spirit that produced these men also provoked the first faint intimation of education in Muscovy. The virtual illiteracy of the Church had not precluded it from dominating such educational facilities as existed. In the middle of the seventeenth century the first school in Moscow came into being. It met with much opposition from the stauncher minds of Greek Orthodoxy; all the more so as the first teachers were Ukrainian monks from the Latinized Academy of Kiev, and Latinity spelt national and spiritual damnation. The school was organized largely through the efforts of another Westernized boyar, Rtischchev, at the Andreyevsky monastery near Moscow. The monastery had been founded and endowed by Rtischchev and he himself studied there, his subjects being Greek, Latin, and Slavonic languages, rhetoric, and philosophy. One of the monks on the staff was commissioned to produce a Greek-Slavonic lexicon. In other ways too the school was influential, attracting some of the less hidebound monks and priests of Muscovy.

In 1687, Rtischchev's institute was joined by the Greek-Latin-Slavonic Academy in Moscow, modelled on that of Kiev. It issued from the merger of two hitherto independent schools, where Greek and Latin respectively were taught. This was a compromise between the partisans of Westernism and the traditionally minded. In actual fact the new Academy became as much an instrument of censorship and intellectual supervision as of learning. For example, only those who had been its pupils could own foreign books; only the Academy might teach foreign languages; the Academy too had to approve the requests of all applicants for foreign-language tutors. Foreign scholars in Muscovy also came under the supervision of the Academy. But it was far from being able to fulfil these various tasks, and by the beginning of the eighteenth century it found itself in a state of utter collapse.

The state of Muscovite intellectual life corresponded generally to the almost complete lack of any provision for education. Arithmetic, which in any case was handicapped by the use of Slavonic characters, since Arabic numerals did not come into general use until the eighteenth century, was limited to addition and subtraction. Euclidean geometry was unknown, as were the new astronomical theories of Galileo, Kepler, and Copernicus. Letters were in an equally parlous state.

However, even if every allowance be made for Muscovite cultural and economic backwardness, the fact still remains that certain fermenting influences were at work. And it was these that helped to produce and formulate the pregnant question: what was the place of Muscovy in history?

Muscovite self-criticism was clearly evident in the writings of

Kotoschichin. Far more significant and conclusive were the views of Krizhanitch. Typically, Yuri Krizhanitch was an out-sider. He came from Croatia and was thus a Catholic subject of the Sultan. He was born in 1617 and studied theology at Zagreb, Vienna, Bologna, and Rome, in the Congregation for the Propagation of the Faith. As a Slav, he was eminently suited for missionary activity in Muscovy, which he had made his adoptive fatherland. His first visit took place in 1647; the second some ten years later. He concealed his priesthood lest he be refused entry to Muscovy. He died in 1683 while defending Vienna against the Turks.

Krizhanitch distinguished himself in Muscovy by offering his services to Alexis as translator, scholar, and political adviser. But he aroused suspicion and was exiled to Tobolsk in Siberia for fifteen years, during which period he also received a small salary from the State. This enforced leisure gave him the opportunity to compose his unfinished *Political Considerations*, in which he tried to define the historical role of the new Mus-covy. Much of this was fantasy in the conditions of the mid seventeenth century, but it was all relevant to the future. Kri-zhanitch did not conceal from himself or his readers any of the evil sides of the life he found in Muscovy – the ubiquitous drunkenness, ignorance, official and judicial corruption and bribery, economic and military inefficiency, personal degrada-tion in matters of hygiene, honour, and punishment. All the same, he was one of the first to contrast the spirituality of Mus-covy with the materialism of the West. Krizhanitch looked on Muscovy with the eyes of the first Pan-Slav. He called on Mus-covy to rescue and unite all the stateless Slav people – the Serbs, Croats, Bulgars, Czechs, and Poles – and restore to them their national honour. The Slavs, Krizhanitch argued, were the young people of the future.

But if Muscovy was to realize its mission to the full, it must guard against the danger of losing its national identity, and succumb neither to the reactionary spirit of the Greeks nor to the heedless progress of the 'Germans'. The Germans, with their superior civilization were the more formidable danger, and had already greatly humiliated the Slavs. On the other hand, in the interests of its mission, Muscovy must reform itself through the introduction of Western culture and enlightenment – translate foreign works on agriculture and trade, develop a mercantilist trading policy, organize its artisans into guilds and its merchants into self-governing corporations, and generally create a system in which every subject of the Tsar would serve the State.

All this, had he known of it, would have been heady stuff to Peter the Great, for Krizhanitch's *Considerations* strongly resembled his own policy. As it is, however, they are even more important as the first attempt to give Muscovy a new *raison d'être* in the changed circumstances produced by the impact of the West.

CHAPTER 7

The Age of Peter the Great

'THE Russian government is an absolute monarchy tempered by assassination,' wrote a nineteenth-century French observer, the Marquis de Custine. A seventeenth-century version would have read: 'tempered by upheavals in the reigning family'. Alexis was twice married – first to Maria Miloslavsky and then to Natalia Naryshkin. For a few years after his death in 1676 there was something of a repetition of the feuds that had disfigured the Kremlin during the minority of Ivan the Terrible – family against family, clan against clan.

Theodore, one of the thirteen children born to Alexis and Maria, succeeded his father on the throne. Under this régime, the Naryshkins, their supporters, and Natalia's eldest son, the four-year-old Peter, were ejected from power and pushed into the background. But in 1682, Theodore died childless. Peter, now ten, was at once plunged into the vortex of a bloody struggle for power. He was proclaimed Tsar; the claims of his half-brother Ivan, an epileptic and almost blind, were passed over. Barely a fortnight later, the *streltsy* invaded the Kremlin, thirsting for the blood of the Naryshkins. There were drunken orgies, there was wild clamour, and in the midst of it all, many of Peter's family were done to death. Peter looked on very calmly, it was said, as the *streltsy* cut up the bodies of their victims in the courtyards of the palace.

In the upshot he was proclaimed joint Tsar with Ivan, under the Regency of Sophia, Ivan's sister. Peter did not become sole ruler in his own name until 1696. By then Sophia had been overthrown, and Peter's mother and Ivan had both died.

The events of Peter's youth may well have fostered an urge to have done with Moscow and all the dead weight of its Byzantine past, its ignorance and inefficiency, its internecine disputes, its complacency, its uncouthness and hostility to change. Be that as it may, Peter was certainly the most venturesome and vigorous of all the Tsars in his efforts to create a Russia (as Muscovy was renamed in this reign) able to rank with the West. It is this that made him not only the most unpopular of all the Tsars during his lifetime but also the most controversial afterwards. Was he the anti-Christ in person? Did his reforms wrench Russia from its natural path and force the country into an alien mould? Was he a calamity for his country or its saviour? Did he introduce a split in Russian culture, for ever afterwards setting a Westernized upper class against an unregenerate mass of peasants? Then there is the ambiguous verdict of Pushkin's poem *The Bronze Horseman*: does one admire above all the Tsar's vaulting ambition and creative energy, or does one give one's heart to the unnumbered victims of his policy? In all these ways the phenomenon of Peter has dominated the Russian imagination.

Today Peter is most impressive as one of the earliest examples of what Toynbee has called '*homo occidentalis mechanicus neobarbarus*'.[1] In similar terms, Peter's efforts have been described thus by Stalin, the man whose own historical role is most clearly analogous:

When Peter the Great was confronted with the more advanced countries of the West, and feverishly went about building factories and mills to supply his army and improve the defence of the country, it was a peculiar attempt to jump out of the framework of backwardness.

Peter was born in the Kremlin in 1672. At the age of five he began to receive the normal education of a Tsarevich. This comprised the rudiments of reading and writing, elementary arithmetic, and the Scriptures. This came to an end in 1682, with the eclipse of the Naryshkins. Peter and his mother were forced to leave the Kremlin and to take up residence in the village of Preobrazhenskoe, one of the favourite country seats of Alexis. Here Peter was left very much to his own devices. The games and interests that he now developed prefigure very closely the man he was to become.

Foremost was his predilection for military toys and pursuits

1. See W. Weidlé, *Russia: Absent and Present*, English translation, London, 1952, p. 38.

– bows and arrows, cannon, toy soldiers. More serious games followed when Peter, still in his early teens, began to drill his companions into regiments and organize manoeuvres around a fortress built on the River Yauza near Moscow. All the paraphernalia of a miniature army were present – cannon and firearms supplied by the royal arsenal, barracks, stables, uniforms, money. Soon Peter had at his disposal two regular regiments, the Preobrazhensky and the Semenovsky. The officers were all foreigners, drawn from the German Sloboda, conveniently close to the village.

The other great interest in Peter's early life was nautical. He had the great good fortune to stumble across an old, crumbling English sailing-boat preserved at the village of Izmailova. It may perhaps have been a century-old gift from Elizabeth to Ivan the Terrible. Peter later referred to this boat as the father of the Russian fleet. In the heart of landlocked Russia it spurred him on to nautical experiment on near-by ponds and lakes. Then, aided by Dutch sailors, he progressed from sailing to building and could indulge to the full his passion for the navy and the art of seamanship. Not until some years later did Peter actually see the open sea at Archangel and travel far into the White Sea.

These early years, a world away from the oppressive, intrigue-laden air of Moscow, were filled also with the eager acquisition of practical knowledge and techniques of all kinds. Peter learnt to lay bricks, work metal, use a printing-press, wield an axe; he became a skilled carpenter and joiner. A Dutchman, Franz Timmermann, first initiated him into such subjects as mathematics, geometry, the theory of ballistics, the art of fortification, and the use of the astrolabe. All in all, Peter mingled perhaps even more freely with the members of the Sloboda than with Russians. His boon companion, for example, was the veteran Scots soldier of fortune, General Patrick Gordon, born near Aberdeen in 1635 and a mercenary under the Emperor, the Swedes, and the Poles. There was Franz Lefort, a cosmopolitan Genevese adventurer and Peter's drinking associate in many a bout. The Sloboda also supplied Peter's first mistress, Anna Mons, the daughter of a German wine-merchant. Peter's Russian companions might be scions of the aristocracy or obscure nobodies, chosen for their convivial qualities.

But the Sloboda and its inhabitants could never be more than a substitute for the real thing; they could not replace first-hand contact with the West. Hence Peter's desire to see and learn for himself. He set out on his own version of the Grand Tour in 1697. 'Peter Mikhailov', the Tsar's thinly disguised incognito – how do you disguise a man nearly seven feet tall? – travelled with an entourage of some 250 people. The aim was to acquire instruction in the latest methods of shipbuilding and to recruit naval and military specialists. The political aim of organizing a grand alliance against the Turks failed through the reluctance of the Western powers.

The Baltic ports of Riga and Libau gave Peter his first view of the Baltic – thence to Mittau, Königsberg, Berlin, Holland, and England. In all, the party was away a year and a quarter.

No Tsar had ever been outside Muscovy before. Still less had any Tsar done what Peter would do. At Saardam and Amsterdam he and his companions worked every day as prentice shipwrights in the wharves of the Dutch East India Company. In their leisure they visited hospitals, mills, workshops of all kinds, schools, and military and naval centres.

After four months in Holland, Peter went on to England. He sailed in the royal yacht presented to him by William III. This was the real climax of the journey. With his headquarters at John Evelyn's house at Deptford, conveniently near the King's Wharf, Peter was able to study the latest shipbuilding techniques. But he also visited Oxford, London, the Tower, the Mint, and, of course, Woolwich Arsenal. There were other visits to Chatham and Portsmouth for fleet reviews and mock naval battles. All of this, Peter meticulously noted in his diary. The lighter side of the English stay produced damages to Evelyn's house estimated by Sir Christopher Wren at £350. Wren's report noted a ruined lawn, destroyed furniture, pictures used for target practice, walls scratched and smeared.

After some four months in England, Peter set out on his return journey, by way of Holland to Vienna. He had not been

long in the Imperial capital when a letter from Prince Romodanovsky, Peter's governor in Moscow and Minister in Charge
of Flogging and the Torture Chamber, urgently summoned him
back – the *streltsy* were in revolt.

The secret police had already uncovered one plot against
Peter led by the *streltsy*. This was shortly before he left for the
West. The ringleaders were tortured and executed. The second
plot had wider ramifications. It drew its strength from the
streltsy's awareness that their power was waning. To their
perennial grievances over arrears of salary had been added the
humiliation of virtual exile to Azov and the south-western
frontier. Hoping to restore Sophia, and inspired by rumours
that Peter had died abroad, a number of regiments marched on
Moscow in the early summer of 1698. Gordon's troops defeated them. But not until Peter's sudden and unexpected return was full punishment meted out.

It was of a scale and type that Moscow had not seen for many
a long day. Day after day, in September and October of that
year, and again in January 1699, men were flogged with the
knout, broken on the wheel, roasted over a slow fire. Report
had it that Peter himself wielded the executioner's axe, displaying the strength needed to decapitate a man with one blow. His
favourites, Menshikov and Romodanovsky, certainly did so.
Some 1,200 men died in this way. Their corpses, often muti-
lated, were left for months on show to the public.

It is no coincidence, perhaps, that this first merciless
onslaught on a characteristic feature of old Muscovy was accompanied by another, though lesser, attack – this time on Muscovite beards. Aided by his court jester Turgenev, Peter in person,
the very day following his return from the West, shaved off the
beards of his principal boyars and also enforced the wearing of
Hungarian or German dress as distinct from the long-sleeved
topcoats worn hitherto. Every Russian, except for the clergy
and the peasants, had to comply with the new laws on pain of
a fine. This effort to Westernize the outward appearance of
Peter's subjects had little success outside the court and military
milieu. Most Russians preferred to pay the fine rather than lose
their beards.

In those early years after the return from the West, Peter also set about reforming local government and finance, replacing the *voevoda* system by elected elders and reorganizing the collection of certain direct and indirect taxes. Most of his ideas came from Holland.

He also devoted these years to the expansion of the fleet and the creation of a new army. An Admiralty was set up, and ship-building feverishly carried on at Voronezh, in which many foreigners recruited from the West were engaged. One vessel, a sixty-gun warship, was designed by Peter himself and built entirely by Russian shipwrights. 'Companies', compulsorily financed by the Church, the merchants, and the larger land-owners, unwillingly supplied the cash.

An army of sorts was raised by levies on the landowners and monasteries. Domestic serfs, slaves, and semi-attached retainers were taken for preference, so that agricultural production would be affected as little as possible. At first, the officers were foreigners and the conscripted gentry.

THE GREAT NORTHERN WAR

Peter's foreign policy barely differed from that of the sixteenth- and seventeenth-century Tsars. As ever, it was a fight for outlets to the seas. At the end of the seventeenth century the struggle for the Sea of Azov, the Black Sea, and the shores of the Caspian was added to the traditional and perennial Baltic struggle.

Russia was in no position to move simultaneously in the north and south. In July 1700, a number of minor campaigns against Turkey came to an end. Russia secured Azov and Taganrog, the right of pilgrimage to Palestine, the abrogation of the duty to make any further 'gifts' (tribute) to the Crimea, and the right to maintain a resident minister at Constantinople.

Peter first struck northwards, where Sweden was the enemy now, not Poland. Ever since the Thirty Years War, Sweden had been the master-power in northern Europe. Its hold on Finland, Karelia, Ingria, Estonia, Livonia, Pomerania, and Schleswig-Holstein made Sweden the Baltic power *par excellence*. Peter's

allies in his challenge were Denmark and Poland. The Poles
attacked first, followed by the Danes. The very day he heard of
the signing of peace at Constantinople, some five weeks after
the event, Peter joined in. On that same day, too, the Danes
were knocked out of the war. This was an inauspicious beginning. But worse was to come.

Peter's forces, all 40,000 of them, under the command of
Eugène, Duc de Croy, Prince of the Holy Roman Empire, were
before the Swedish port and strongpoint of Narva. Suddenly
the Swedish troops appeared, led by the eighteen-year-old
Charles XII, and overwhelmed the Russians in a thick snow-
storm. The Russian troops fled in disarray or gave themselves
up.

Peter was in despair. Peace at any price was his first desperate
cry. But this defeat evoked new strength and resilience, for
Peter naturally made the reform and reconstruction of the army
his first concern. Levy after levy was imposed on the peasants
and townspeople, usually in the proportion of one man per
twenty households. This produced about 30,000 men a year.
Training, drill, and tactics were all revolutionized. Peter
stepped up the production of flint-locks and bayonets, of siege
guns and field-artillery. New ironworks and powder factories
in the Urals were the chief centres of arms manufacture. He
made every effort to attract more and more military specialists
from abroad; and in 1702 issued a decree inviting all foreigners
to Russia, promising them religious freedom and their own law
courts, as well as free passage and employment.[1] At home,
further measures of conscription were imposed on the landowners. To finance all this effort, Peter sequestered the Church's
revenues; all in all, about eighty per cent of the State's income
went to feed the needs of war.

But for all his urging and impatience, Peter also needed time.
Here Charles came to his aid. He refused to follow up his
victory at Narva by a march on Moscow; instead, he moved

1. Jews were a notable exception. Peter's anti-Jewish sentiments were
reinforced by Menshikov's similar aversion. After Peter's death, Menshikov forbade all Jews to enter Russia and brought about the downfall
of Baron Shafirov, Peter's envoy at Constantinople – a converted Jew.

south and involved his forces in innumerable minor campaigns against the Poles. Peter himself kept Charles bogged down in Poland as long as possible by sending Augustus II, the Polish King, reinforcements and money. But at the end of 1706 the Polish front finally collapsed. By the Treaty of Altranstadt a Swedish-Polish peace was concluded. Augustus had to give up his throne and acknowledge Charles's nominee, Stanislas Lesczynski, as the new Polish king.

In the meantime, however, Peter had also to some extent been able to strengthen his position, especially to the north. He had retaken Ingria, the port of Dorpat, and stormed Narva (1704). In 1703 he had founded St Petersburg (now Leningrad) near the mouth of the Neva River and constructed the fortress of Kronstadt to protect it from the open sea.

But none of this could be of great avail against Charles, who was now freed from the incubus of the Polish campaign. Peter had already made several attempts to secure a mediated peace, on the sole condition that he retain St Petersburg. Their failure left him all the more exposed to Charles's imminent attack. In addition, it was at this time that his repeated levies, programmes of forced labour, and oppressive taxation provoked uprisings in the Russian interior. There was one revolt at Astrakhan in 1705. Two years later the Don Cossacks rebelled, led by their *hetman* Bulavin. The movement rapidly spread to a mass of peasant refugees from serfdom in the Don country. From there it engulfed the Voronezh area and the workers conscripted to build the Volga-Don canal. Eventually, Bulavin's uprising dominated a vast area between the Volga and the lower Don. It also set off sympathetic rebellions among certain of the oppressed non-Russian nationalities – the Bashkirs, Tartars, Tcheremis.

The suppression of these revolts urgently demanded the transfer of troops needed in the West; even so, the task was mercilessly carried out. Foreign observers were saying, with evident truth, that had Peter's campaign against the Swedes failed, he would have had to face a revolution.

In 1706, Charles began to march eastwards into Poland. The Russians systematically retreated, avoiding pitched battles as far as possible, and carrying out a scorched-earth policy. In July the enemy crossed the Dnieper and took Mogilev. Would Charles then strike northwards at Novgorod and Pskov and make his ultimate goal St Petersburg? Or would he aim directly eastwards at Smolensk, the classical key to Moscow, only some sixty miles away?

At first, Charles dismissed the Livonian area which had been fought over too much already to withstand further campaigning. He resolved to aim at the Russian centre – Moscow. The capital hastily prepared for the imminent siege. But the central area had also been thoroughly scorched. Charles changed his plan. He decided on a detour by way of the Ukraine. Here there would be no lack of food supplies for his troops; moreover, in the person of Mazeppa, the *hetman* of the Ukraine, he had a secret ally, who, to Peter's fury, chose this precise moment to climb on Charles's band wagon.

But this move southwards proved fatal. Not only did Mazeppa fail to carry with him the bulk of the population, but Charles had moved farther from his supply and baggage train, which was slowly advancing southwards from Riga under the command of General Löwenhaupt.

The two Swedish forces were isolated. At the small town of Lesnaya, to the south-east of Mogilev, the Russians attacked Löwenhaupt's army and inflicted a heavy defeat. The vast bulk of the supplies of arms and munitions eagerly awaited by Charles had to be destroyed. This was in October 1708. An unusually severe winter followed. The Swedish forces, not much more than 20,000 by now, suffered from the climate and shortages of food and arms. The climactic battle came on 27 June 1709 (8 July new style). Charles besieged the small fortress-town of Poltava – and suffered a crushing defeat. The Swedish survivors retreated to the Dnieper. But their boats were burnt and they fell before the onset of Russian cavalry. Charles and Mazeppa fled to Turkish territory. In a phrase, Peter crystallized the meaning of the victory: 'Now the final stone has been laid on the foundations of St Petersburg.'

But the 'most glorious victory' of Poltava was far from ending the war. The Great Northern War, as it came to be called, dragged on with intermissions for another twelve years. But at least Russia was freed from the threat of invasion. This was perhaps its most important immediate consequence. By the same token, it gave Peter the opportunity to exploit the Swedish collapse so that he could consolidate his position to the west and north. There was a radical transformation in the Russian diplomatic position.

Peter could overrun Livonia and Estonia and control the whole Baltic coast from the Western Dvina to Viborg; he could reinstate Sigismond on the Polish throne; he could fulfil the dream of Ivan III and Ivan the Terrible and introduce Russian men-of-war into the Baltic, an unprecedented phenomenon; he could marry his nieces to the rulers of Mecklenburg and Courland and dabble more or less successfully in North German politics; he could even aspire to a marriage between his daughter Elizabeth and the young Louis XV of France; he could conquer Finland and raid the Swedish coast.

The beginning of the end of the Great Northern War came in 1718; direct Russo-Swedish negotiations opened on one of the Aaland Islands. They dragged on inconclusively for nearly eighteen months. A further year and a half later, the final peace was concluded at Nystad. Russia received the territories of Livonia, Estonia, Ingria, part of Karelia, with the city and district of Viborg, and the islands of Ösel and Dagö. In return, Russia bound herself to restore Finland to Sweden, to pay a compensation of two million Dutch thalers, and to refrain from interference in Swedish internal affairs. These terms signalized nothing less than the emergence of Russia as one of the great powers of Europe.

The effect of this was most marked in the diplomacy and politics of Northern and Western Europe, but it by no means exhausted Peter's expansionist policy, especially to the south. This began with a sad failure. In his second Turkish war, two years after the battle of Poltava, the Russians were defeated and had to yield Azov to the Turks, losing all the gains of the campaign of 1700. Peter had hoped to exploit the Orthodox sympathies of the Balkan Slavs – he was the first Tsar to take this revolutionary step – and to pose as their protector against the infidel Muslims. But the response was indifferent, and this helped to account for the eventual Russian defeat.

Peter was more successful in the Far East and central Asia. He was able to develop the trade in silk and furs with China, while leaving unchanged the political relationship established by the treaty of 1689. In the far north he advanced the Russian frontiers to Kamchatka and annexed the Kurile Islands in the Pacific. In central Asia his efforts to subdue the two Muslim khanates of Khiva and Bokhara failed.

Peter's most noteworthy successes were in Persia and along the shores of the Caspian. He sent his first envoy, Volynsky, to Ispahan in 1715, hoping to expand trade with Persia, explore the possibility of a route to India, and enforce a rerouting of the Armenian silk trade through St Petersburg, cutting out the route through Turkey. Volynsky did indeed conclude a trade treaty that opened up Persia to Russian merchants. More important, he reported that the country was in a state of collapse; nothing more than a small detachment would be needed to carve it up. For the moment, preoccupation with the Great Northern War inhibited Russian aggression. But in 1722, on the pretext of aiding the Shah to restore order in his territory, Peter set sail down the Volga to Astrakhan. Derbent soon fell to the Russians, then the port of Resht, and in 1723 Baku. In the eventual peace, Russia retained these ports, and also certain Persian provinces along the Caspian southern and south-western littoral. As ever, Peter had grandiose ideas for developing these areas. He would build fortifications, deport the Muslims to make way for Christian settlers, carry out mineral surveys. . . . But it came to nothing in the end. Soon after his death, the newly acquired provinces had to be abandoned provisionally, largely because of the unhealthy climate.

THE REFORMS

For most of Peter's reign, Russia was at war. This very largely determined the nature, scope, and success of the Tsar's activities as a reformer. It is in a way a tribute to the intractability of the Russian people that in every century the task of yoking the nation to the State had to be taken up anew. Under the impact of war and an unremitting struggle for existence, the same vision of a people devoted to the State had inspired the rulers of the fifteenth, sixteenth, and seventeenth centuries.

But to none was this vision more of a reality than to Peter. He shrank from no measure that seemed calculated to advance his purposes, however great the opposition he met with. 'The Tsar pulls uphill along with the strength of ten, but millions pull downhill,' wrote Pososhkov, an early apologist for the Petrine régime. To drag these millions with him, Peter had to deploy a widespread system of spying and informing, in addition to the punishments traditional in Muscovy. He made his subjects, particularly the peasants, pay a heavy price for their country's rise to the status of a great power.

Money engaged Peter's attention from the start. How could he cover an ever-mounting military expenditure – 2·3 million roubles in 1701, 3·2 millions in 1710, over 4 millions in 1724? There were a number of expedients to hand, all of which Peter made use of. He confiscated the income of the estates of the Church; he debased the currency; he introduced any number of indirect taxes, which were often suggested by a special corps of 'profit-makers'. There were taxes on beards, coffins, bee-keeping, bath-houses, knife-grinding, articles of clothing, the sale of salt, the weddings of non-Russian tribesmen. But by 1710 these measures had proved ineffective. This produced a transformation in the incidence of direct taxation. This had hitherto been based on a tax on households as determined by the census of 1678. In 1710 a new census was ordered in the confident hope that the result would show an increase. But it actually revealed a decline of about twenty per cent over the whole country, even after making full allowance for the inefficiency of the enumeration and the widespread evasion practised by the taxable unit. But for the moment the old system was retained. In 1719, however, after a further census had revealed an increase in the number of male peasants, the transition was made to a poll-tax on the individual male. The penalty for evading the census was death. The new tax succeeded both in increasing revenue, bringing in more than fifty per cent of the budgetary income in 1724, and in extending the cultivated area. The peasants had to cultivate more land in order to pay the tax. The new system lasted until 1886.

Peter's industrial and economic policy ran on more traditional lines. He encouraged the industrialization of the country on the lines already laid down by the Romanovs of the seventeenth century. But Peter emphasized more strongly than ever before the manufacture of munitions and arms, and of textiles for the clothing of his soldiers and sailors, in order to be independent of Western supplies. The production of paper – indispensable in a bureaucratic régime – also went forward, so that by 1723 Peter could decree that only Russian-made paper be used in the departments of state. At the end of the reign, Russia had some 200 factories, some of which employed more than 1,000 workers. Among the largest factories were those that produced wool, linen, silk, leather, cotton, and ornaments.

Mining and metallurgy also progressed. The Ural iron deposits were the scene of the most notable advance in extractive technique and processing. Between 1695 and 1725 fifty-two new ironworks were opened, of which one-quarter were in the Urals. Altogether, the latter accounted for about forty per cent of all Russian production, equally divided between State-owned and privately owned works. In not much more than a quarter of a century, Russia had come to lead the world in the production of iron. It was Russia which supplied the major part of England's iron imports.

The general development of Russian manufactures and mining depended on State aid, forced labour, and protective tariffs. There was indeed a preponderance of merchants and mercantile capital among the companies formed, at Peter's urging, to develop the country's resources; but State subsidies were also essential in view of the shortage of capital. The State similarly helped to overcome the shortage of labour. It conscripted all sorts of vagrants, unemployed and marginal individuals, such as prostitutes and orphans, for service in the factories. In less populated areas the problem would be overcome by forcing peasants from State lands into factory service. Runaway peasants working in a factory could not even be reclaimed by their owners. Such was the emphasis Peter gave to industrialization. Finally, the State not only helped to subsidize and to man new enterprises, it also gave them the benefits of a high tariff, the right to import machinery and raw materials freely, and exemption from taxation.

Financial and industrial policy had only an indirect effect on the status of the various classes. Peter's social policy, on the other hand, intentionally produced profound changes in their duties and interrelationship. As ever, the aim that overrode all others was to organize the population so that it could best fulfil its duties to the State. This principle of universal and compulsory national service had nothing new about it. But Peter systematized, refined, and intensified its application.

The position of the landowning *dvoryantsvo* – the body of the nobility – underwent the greatest change. They were first forced to undergo instruction in some branch of practical knowledge, either abroad or in certain 'mathematical' schools that Peter established in St Petersburg. The penalty for failing to complete the course was enforced celibacy! At fifteen, the young *dvoryanin* had then to choose between State service in the army, the navy, or the bureaucracy. Russia had no nautical tradition and the *dvoryanin* did not take kindly to the sea; the latter career was the most favoured. It was safer, less arduous, and more remunerative. To preserve a balance, therefore, personnel were rationed, only one in every three young men being allocated to the bureaucracy. Those who 'chose' the army as their career served in one of the three regiments of Guards. They had to work their way up from the lowest rank before being rewarded with commissions in other regiments, or, if their connexion and status warranted it, with a commission in the Guards themselves. It was this special status that later in the eighteenth century gave the Guards their importance as kingmakers. A special official, the Heraldmaster, controlled the *dvoryanin*'s career. He maintained a register of each family, ensuring, on pain of the severest penalty, that no leadswinger, slacker, scrimshanker, or defaulter evaded his duty.

Peter reinforced the threat of punishment by a flanking attack that struck at the very roots of the ancient Muscovite law of inheritance. Estates had hitherto been divided equally among the sons of the deceased. This had led to much subdivision into unprofitable units, with consequent loss of tax-paying capacity and also waste of manpower. To overcome these twin disadvantages, Peter issued an unprecedented Entail Law in 1714. He drew largely on English practice in primogeniture. An estate could henceforth be bequeathed to one son only, chosen by the owner. The remaining children, Peter hoped, would be forced off the land and into State service. But the new law, more at home, it has been well said, in an aristocratic England than a bureaucratic Russia, did not long survive Peter's death and was abolished in 1730.

He had much more success with the institution of the 'Table of Ranks' of 1722; this lasted right up to the Revolution of 1817. It produced a fundamental change in the structure of the class of nobles and gentry, and was the Russian version of Napoleon's ideal of *'une carrière ouverte aux talents'*. If the *dvoryantsvo* had to serve, then it was only logical that those who served should be, or should become, members of the *dvoryantsvo*. What Peter did was to classify into fourteen parallel grades all the ranks in the army, navy, and bureaucracy. Those who reached the eighth rank from the top in the bureaucracy were granted ennoblement. In the army and navy, ennoblement was earned on reaching the lowest commissioned rank. Service in both military and civilian posts began in the lowest grade; promotion depended on merit and seniority. The whole scheme was much more in keeping with Russian tradition than many of Peter's other reforms. The nobility had never formed a closed corporation or caste. Now hosts of newcomers could enjoy the privilege of owning serfs, and also exemption from the poll-tax. There was, of course, opposition to the reform on the part of the older noble families, but it was not in their power to arrest the influx of parvenu nobles. On the other hand, it is also true that social connexions still had the edge on merit in securing promotion. The real importance of Peter's 'democratization' of the nobility lay, it seems, in creating for society an eminently bureaucratic framework.

The first emergence of the *dvoryane* in the fifteenth and sixteenth centuries had been marked by their intensified control over the peasant. Their further ascent in the early eighteenth century had the same result. The serfs in Peter's time made up about ninety per cent of the whole population of some twelve million. This total was enlarged by including the class of slaves, who were thus made liable to military service and to the poll-tax. The State peasants, monasterial retainers, and dependents were also subjected to conditions closely resembling serfdom. The State laid further burdens on these unfortunates. There was the poll-tax; there was forced labour in the new mines, factories, and ironworks; the construction of St Petersburg on marshy, Finnish swampland demanded the lives of thousands of labourers; the digging of canals and conscription claimed others.

Those who stayed at home by no means escaped. On the contrary, the State subjected them ever more tightly to the landowner. In 1721, Peter issued a decree denouncing the sale of serfs 'separately, like cattle, a practice which is not supported anywhere else in the world'. But he introduced no penalty for its infringement and could only suggest that whole families be sold; as ever, the landowner remained sole ruler of every aspect of the serf's way of life. Moreover, the introduction of an internal passport system obliged every serf to secure his owner's written permission before being able to leave his village. This system lasted until the Revolution, and even survived it in a somewhat changed form.

The merchants and the urban population, on the other hand, derived some benefit from the reform. The municipal autonomy enjoyed by Riga and Revel was Peter's inspiration. He divided the town population into three main groups – the 'first guild' consisting of the upper bourgeoisie of wealthy merchants, professional men, the 'second guild' of small merchants, handicraftsmen, artisans, and the 'common people' of hired labourers and the poor generally. The two guilds jointly elected a magistracy responsible for all the town's affairs, on which, however, only the members of the first guild could serve. Peter also attempted to establish true guilds of craftsmen of the type familiar in the Western-European town, but this failed.

On the whole, one may say that Peter had more success in remodelling the organs of central and provincial government than in anything else. But before he found an adequate answer to the problems confronting him he took many a false step. Moreover, owing to the prevalence of bribery, peculation, overlapping functions, ill-defined powers, and the sheer physical difficulty of controlling vast areas suffering from poor communications, Peter had constantly to grapple with the problem: *quis custodiet custodies?* At first he took the path of decentralization. He created eight provincial governments (to which he later added two more), with very extensive powers. But this left a hole at the centre, so in 1711, when he left for the second Turkish war, Peter established a senate of nine members to supervise the work of the provincial governments, to function as the supreme court of justice, and above all, to ensure that all taxes were efficiently collected. A Guards officer was deputed to attend its meetings and to supervise the senators' conduct – no easy matter if the punishments to which they were subjected is any criterion. In 1715, for example, two senators were flogged, and had their tongues branded and their property confiscated. In 1722 a general-procurator replaced the Guards officer as Peter's 'eye'.

The Senate lasted until 1917. Less enduring, but still of fundamental governmental importance for a century, was Peter's introduction of nine colleges, based on Swedish and Danish models. They completely replaced the old system of Prikazy. Each college had a functional jurisdiction – army, foreign affairs, revenue, mining, and manufacture, for example – and was governed by a Russian president, assisted initially by a foreign-born vice-president, and a board of eleven members.

The old *monastirskii prikaz*, which had hitherto enjoyed a certain limited power over the administration of the Church, was completely revamped into the Holy Synod. This replaced the Patriarchate and took over the latter's responsibility for all spiritual and ecclesiastical matters. All the property of the Church also came under the Synod's control. The Synod was in fact a department of state, a college subordinate to the Senate and conducted on the same lines as the other colleges (except that its members were all clergy). The same spirit of making the Church serve the State persuaded Peter to restrict entry into the monasteries and to divert the activity of the monks and nuns into such socially useful paths as the care of the sick and orphans, and artisanry. The Church, weakened by the effect of Nikon's reforms, gave little opposition to its diminished status.

Where provincial administration was concerned, Peter's over-faithful adoption of Swedish models led to an artificial system of fifty provinces covering the entire Empire. Its main intent was not only to achieve organizational tidiness but also to separate administration *qua* administration from the administration of justice. Thus the fifty provinces expressly did not coincide with the eleven judicial districts. Peter further complicated the position by removing from the purview of the provincial authorities the collection of the poll-tax and the selection of conscripts. This was tantamount to military rule in the countryside, in view of the importance of these two requirements of State. But the whole system produced so much tension between the rulers and the ruled and was so ineffective and unwieldy that it did not long survive Peter's death. The attempt to separate justice from administration ran clean counter to Russian tradition and to the spirit of indivisible autocracy as it had developed ever since the days of Ivan III.

The acquisition of the new Baltic territories and the gradual Russian expansion into the Ukraine posed administrative problems of their own. The German ruling class of landowners in Estonia and Livonia was left undisturbed in the enjoyment of its special privileges, and the position of the Lutheran Church confirmed. In the Ukraine, on the other hand, consistent politico-economic Russification was the order of the day. At Kiev a Russian governor and two Russian regiments wielded power; and Russian nobles would receive grants of land in the Ukraine for particularly distinguished service. In the case of the peoples with distinctly less developed culture – the Bashkirs and Tartars of the Urals and the Volga, for example – the policy of Russification could be applied more harshly. No non-Orthodox Russian could own serfs. Such peoples were also exposed to the full force of Orthodox missions. Consequently, Tobolsk became the seat of a Metropolitan as early as 1700.

The same spirit of enlightenment, secularization, and practical utility informed Peter's educational and general pedagogic policy. He forced Russia to study – or rather, certain Russians to study certain subjects. The publication of about thirty decrees devoted to education in a broad sense testify in themselves to Peter's perennial concern. But it was an uphill struggle and not always successful.

When Peter ascended the throne he found only one school in existence, the Moscow Academy, apart from two theological academies in Moscow and Kiev respectively. This was plainly insufficient provision for the hosts of new experiments and specialists that the new State demanded. Foreigners supplied much of the trained personnel, but this could be no more than a stopgap. In any case, they fulfilled in the main an executive role and had little influence on policy. Russians, trained abroad in navigation, seamanship, gunnery, economics, languages, engineering, could also help. But the bulk of a trained ruling class must obviously be produced inside Russia itself. Peter, concentrating on the needs of the army and navy, first established a number of schools in Moscow and St Petersburg for future officers in the two services. Their curriculum was limited to mathematics, navigation, gunnery, and military engineering. The schools acquired their pupils by conscripting the sons of the *dvoryantsvo*. But even this method hardly produced sufficient pupils.

Peter's plan for the introduction of an empire-wide system of elementary education had even less success. In 1714 each province was ordered to establish two 'cipher' (mathematical) schools. The Admiralty bore the cost and the provincial authorities supplied the pupils, again on the basis of conscription, this time primarily from non-noble families. Here also, however, the pupils dwindled away, and the schools with them. In 1722 the number of schools totalled forty-two, with some 4,000 pupils. Three years later the totals were twenty-eight and 500 respectively. By 1744 the whole system had collapsed and was merged with a network of army schools run by and for the local military garrisons.

Part of the reason for this failure was the success of the Church from 1721 onwards in organizing a parochial school system for the sons of the clergy. This drew off many of the pupils who would otherwise have attended the 'cipher' schools. Only one school of a non-vocational type existed. Pastor Glück, a Lutheran from Marienburg, founded this establishment in Moscow and also drew up its curriculum – Oriental and European languages, literature, ethics, philosophy, equitation, etiquette, and decorum. The teachers were all foreigners, but in ten years its pupils had dwindled to five and the whole institution was disbanded.

Yet more grandiose in appearance, but in actual fact a permanent feature of Russian intellectual life until the Revolution, was Peter's foundation of the Academy of Sciences. This owed something to the inspiration of Leibniz, whom Peter met twice, and to the examples of the Königliche Preussische Akademie in Berlin and the Royal Society in London. At first, its professorial fellows and their students had all to be imported from Germany. But this did not later inhibit the growth of a truly Russian institution.

To Peter's educational activity in the widest sense of the term there was no limit. He remodelled the calendar, adopting the Julian system in 1700 so that Russia no longer reckoned from the creation of the world but from the birth of Christ; he simplified the old Cyrillic alphabet, reducing the number of characters and approximating the remainder to Latin models; he founded the first Russian newspaper, and urgently promoted the translation of foreign works – those which were overwhelmingly secular and technological in tone; he inaugurated the first public Russian theatre, with performances on the Red Square in Moscow; he forcibly ended the seclusion of women; and he instituted *assemblées*, where the nobility of both sexes were forced to acquire the rudiments of *bon ton* by entertaining one another with gossip, dancing, and refreshment.

But for all Peter's urge to have done with old Muscovy and to create a modern Russia, no consistent policy of Westernization ever existed. Foreign policy was largely traditional. As far as the Baltic was concerned, Peter openly saw himself as the successor to Ivan the Terrible, whose portrait flanked his own in the triumphant procession after the battle of Poltava. At home, too, much of Peter's policy was built on the foundations of the seventeenth century. The army, the manufactures. education, the employment of foreigners – all these features of Peter's reign signified nothing new. There is thus no single antithesis between the 'organic' life of Muscovy and the 'inorganic' life of Petrine Russia, though such antithesis is implicit in the many attacks on St Petersburg as the epitome of the new régime – for example, in Dostoevsky's condemnation of the new capital as 'the most abstract and artificial town that exists'; or in Khomyakov's description of 'the dead beauty' of St Petersburg: 'a city where all is stone; not only the houses but the trees and the inhabitants'.[1]

But if Peter did not break with the past, it is also clear that he did inaugurate some kind of new order, some kind of new departure. Furthermore, it is obvious that this cannot be equated with the dichotomy of Russia and Europe. A law of Entail *à l'anglaise*, a collegiate system on the Swedish model, an Academy of Sciences taken from Berlin, military and naval tactics from Great Britain and Holland – all these, and much more, do not in combination amount to the Westernization of Russia. All such imported ideas and institutions had a connotation very different in Russia from what they had in Europe.

Ever since the time of Ivan III and Ivan IV there had been a consistent Russian attempt to catch up and surpass the more technically advanced countries of Western Europe. And it was to this attempt that Peter's reign came as a partial climax and a logical sequel. It is implicit in Peter's remarks to the Swedish generals taken prisoner after Poltava: 'Gentlemen,' he said, 'it is to you we owe it.' Russia used the West in order to overcome the West. From Ivan the Terrible to Peter the Great – and beyond – the West appeared first and foremost as a system of power, the adoption of which would serve as the sole means to ensure Russian survival. It is in this sense that Peter was a West-

1. For Khomyakov, see p. 153.

ernizer and in this spirit that he strove to modernize Russia and make it a part of the West.

The most profound feature of our historical physiognomy is the absence of any spontaneity in our social development [wrote Tchaadayev]. Every important feature in our history is imposed on us from above, every new idea is imported. Peter the Great found at home in Russia only a blank sheet of paper. With his powerful hand he wrote on it 'Europe and the West'; since then we have belonged there. . . .

But who is this 'we'? Who belonged to Europe and the West and who did not? In the early stages of Peter's reforms there was undoubtedly a certain incongruity in turning Muscovites into Westerners. Western clothes sat comically on Muscovite shoulders, sometimes literally so. For example, in 1701 the British Ambassador to Turkey drew attention to the contrast between old and new:

The Muscovite Ambassador and his retinue have appeared here so different from what they always formerly wore that ye Turks cannot tell what to make of them. They are all coutred in French habit, with an abundance of gold and silver lace, long perruques and, which the Turks most wonder at, without beards. Last Sunday, being at mass in Adrianople, ye Ambassador and all his company did not only keep all their hats off during ye whole ceremony, but at ye elevation, himself and all of them pulled off their wigs. It was much taken notice of and thought an unusual act of devotion.

This cheek-by-jowl juxtaposition of the old and the new gradually gave way, however, to a more profound influence, even if in absolute terms it never did go very deep. With all its faults, the culture and the intellectual world of Muscovy had been homogeneous, at least, until Nikon's reforms forced a breach. From the Tsar to the last *muzhik*, however much their differing socio-political roles might set them at loggerheads, there was a basic unity in values, outlook, and prejudice.

It was this unity that Peter's reforms destroyed. In its simplest and most obvious form, a bewigged and clean-shaven *dvoryantsvo* and bureaucracy, dressed in Western style, pursuing a career in some form of State service, and living apart from its hereditary estates, confronted the mass of bearded peasantry. The beard was far more than a symbol. It stood for two worlds of the mind. A tiny minority, educated in Western ideas but as yet by no means permeated with them, lived a very different life from the 'dark mass'. Two cultures lived side by side with the minimum of interrelationship or interconnexion. The mass, steeped in poverty and illiterary, clung with ingrained conservatism, broken only by spasmodic elemental upheavals, to its traditional religion, way of life, dress, and morals. The alienated upper stratum spoke an entirely different language, thought in different terms, and pursued different ideals. To quote a recent student: 'On the ideological foundation of the traditional, religious Tsarist charisma, Peter ... erected a purely secular superstructure formed on the model of Western enlightened absolutism.' [1] This division was a fateful element in Peter's legacy to the Russian future. It needed another, and even more fundamental revolution – the Bolshevik – to overcome the split.

1. Emanuel Sarkisyanz, *Russland und der Messianismus des Orients,* Tübingen, 1955, p. 154.

CHAPTER 8

Russia of the Nobles

PETER THE GREAT died in 1725 from stone and strangury. He was unable, without great pain, to retain or discharge his urine. This was a consequence of syphilis, probably contracted in Holland and aggravated by drunkenness. He left no successor. From a combination of personal and political motives, he had already caused the death of his son Alexis, and had excluded Alexis's son from the succession. By a decree of 1722, Peter claimed the right to nominate his successor in much the same spirit that had animated Ivan III 'to whom I will, to him I shall give my throne'. But when the time came, Peter was unable to speak. One of his daughters waited at his death-bed for the word that never came.

Given the tensions and the conflicts that Peter's reform programme had let loose inside Russian society, a period of reaction and disorder would have been normal. But what actually happened exceeded all precedents. Between Peter's death and Catherine's accession in 1762, the throne had no less than six occupants: Catherine I, Peter II, Anna, Ivan VI, Elizabeth, and Peter III.

Those thirty-odd years were the palmiest of palmy days for the favourite, the intriguer, the lover, the instigator of the midnight *coup d'état* and palace revolution, the politician of the bedchamber, the king-making Guards, and the assassin. Writing in 1741, Daniel Finch, the British Minister in St Petersburg, most accurately summed up the confusion of the Petrine system:

> After all the pains which had been taken to bring this country into its present shape ... I must confess that I can yet see it in no other light than as a rough model of something meant to be perfected hereafter, in which the several parts do neither fit nor join, nor are well glued together. ...

But amid the welter of changing rulers two great facts stood out. First, there was no weakening of Russia's position as a great European power. On the contrary, in all the wars and diplomatic transactions of the eighteenth century, such as the carve-up of Turkey, the Seven Years War, the elimination of Poland, Russia increasingly brought its weight into play. Second, inside the country, the incessant struggle for the throne diminished the prerogatives and power of the autocracy, and permitted the efflorescence of the nobility. What the autocracy lost, the nobility gained.

The background to this development indicates the transition through which Russia was passing in the eighteenth century. It was probably a question of war. Roughly speaking, up to and including Peter's reign, Russia's wars had been wars of existence. At least, they had involved the concentrated exertion of the population's energies and also determined the country's social evolution. But Peter's victories finally did away with the threat of national extinction. There were wars in the eighteenth century, of course, but they were waged from a position of strength, usually not on Russian soil, and altogether lacked the desperate do-or-die quality.

Thus there was scope for relaxation all round. And of this the *dvoryantsvo* was not slow to take advantage.

It had two main aims: to shake off the burden of national service imposed on it by Peter, and to strengthen its position *vis-à-vis* the serfs. It largely succeeded in both. This does not mean, of course, that the Petrine absolutist and bureaucratic type of state was dismantled overnight; or that there were not significant divisions of interest among the nobility. But the general drift of events is clear enough.

The first open indication of the way the wind was blowing came as early as 1730. The Supreme Privy Council, a body of would-be oligarchs that had replaced the Senate as the highest judicial and administrative organ, decided to offer the vacant throne to Anne, the widowed Duchess of Courland and a niece of Peter the Great. But there were conditions attached to the invitation, similar to those imposed on the Swedish crown by the Swedish nobility after the death of Charles XII. Without the consent of the Council, Anne must not marry or appoint an heir, make war or agree to peace, levy taxes, grant senior army or civil rank, bestow estates out of Crown land or dispose of State revenue, deprive a noble of his life or estate without due trial. Anne agreed to these restrictions of power. But on arrival in Moscow for her coronation, she dramatically tore up the list of conditions. Supported by the Guards regiments and many of the lesser nobility, she was able to rout the would-be oligarchs of the Supreme Privy Council. The way ahead lay clear for a widespread expansion of the nobility's prerogatives.

The first to go of Peter's unpopular measures was the Entail Law of 1714. At the end of 1730 the *status quo ante* Peter was restored, leaving the nobility henceforth free to subdivide its estates among all the heirs. The next blow at the old régime came the following year, with the establishment of a military academy through which the sons of *dvoryane* were channelled directly to commissioned rank in the army or the bureacracy, thus avoiding the necessity of starting in the ranks. In 1736, the period of compulsory State service was reduced from life to twenty-five years. Furthermore, if a father had two sons, one was altogether freed from service so that he might devote himself to the care of the family estates. (Owing to the Turkish war, this decree did not come into active operation until 1739.)

The short reign of Peter III (January–July 1762) produced a whole series of liberal decrees, culminating in the emancipation of the nobility from State service entirely. The Tsar's manifesto put the nobles' obligations on a purely voluntary basis, empowered them to resign from the army or the bureaucracy except in time of war, to travel freely abroad, and to take service with non-Russian powers. (But those who retired from service were prohibited for life from appearing at Court.) 'Altogether, it is difficult to exaggerate the importance of the edict of 1762 for Russia's social and cultural history. With this single act, the monarchy created a large, privileged, Westernized leisure class, such as Russia had never known before.' [1] Peter also abolished the secret police, prohibited the purchase of serfs

1. R. Pipes, *Karamzin's Memoir on Ancient and Modern Russia*, Cambridge, Mass., 1959, p. 15.

by manufacturers for service in the factories, and secularized the estates of the Church. Thereafter, the institutions of the Church and its various dignitaries would receive specific sums for the maintenance and upkeep from a State-administered college. This followed logically on Peter the Great's abolition of the Patriarchate.

The climax to the whole process of the emancipation of the nobility came in 1785 with the promulgation of their 'Charter'. This document reaffirmed and widened all the special status won in the previous half century – the right to travel abroad, enter the service of friendly foreign states, resign from government service, dispose of property at will. The *dvoryanin* could not be dispossessed of his title, estates, life, or personal status without trial by his peers. If convicted on a serious charge, his hereditary estates would pass to his heirs and not be confiscated in favour of the State, as had formerly been the custom. The *dvoryanin* enjoyed exemption from the poll-tax, corporal punishment, and the duty to billet troops. He could engage in commerce and industry, and exploit on his own account any mineral or timber resources on his estates.

There was also a provision for the formation in each province of a corporation of nobles who would cooperate with the organs of local government by electing certain of the *dvoryane* to administrative posts. But where these functions were not decorative, they were much overshadowed in importance by the centrally organized system of provincial and local government. In effect, the decrees of 1762 and 1785 were an eighteenth-century restatement of the alliance of autocracy and nobility that had first been concluded in the sixteenth century. The Tsar would rule at the centre, the nobles in the countryside. Furthermore, class division among the nobles forced many into government service for economic reasons. In 1777, for example, it seems that as many as thirty-two per cent had less than ten serfs; twenty-seven per cent had from ten to twenty; twenty-five per cent from twenty to one hundred; and only sixteen per cent more than one hundred. Since the nobles, in general, took little part in economic development and regarded their estates and perquisites purely and simply as a means of support, there was some economic incentive to enter State service. But the effect of this too should not be exaggerated. In 1762 there was quite an exodus from St Petersburg, and the following decades saw the growth of a sizeable class of provincial nobles.

As always in such circumstances, the emancipation of the nobles was accompanied by the reaffirmation of their rights over their serfs. All crimes except murder and robbery came within the landowner's jurisdiction. A number of decrees of the 1760s further extended this power to include the right to send criminal or delinquent serfs into Siberian exile as settlers, or to periods of penal servitude in Siberia. To serve in the army was another form of punishment.

Upheavals and revolts were regular among the serfs, both on private estates and in the possessional factories. The landowners often needed to call in regular troops to crush the rebels. The climax came in 1773 with the uprising led by Pugachov. This exceeded in scope and terror any peasant uprising of the seventeenth, eighteenth, or nineteenth centuries. The movement drew its vigour not only from the perennial burdens of the peasantry; but also from the widespread hope that since the nobility had been freed from State service, the serfs would also be similarly emancipated.

Pugachov was a Don Cossack who gave out that he was in fact Peter III. After a nomadic life, in and out of prison, in and out of the Russian army, he raised his revolt in the territory east of the Volga in the autumn of 1773. The provinces of Orenburg and Kazan were its centres. In these vast regions grievances of all kinds lay ready to hand – the dispossessed native tribes, misruled and subjected to forced conversion; the Old-Believers; the forced labourers in the Ural mines and factories; the Cossacks perpetually at odds with the central government; over and above all, the refugees from serfdom.

Pugachov's programme did not differ in essence from that of his predecessors, from Bolotnikov to Stenka Razin and Bulavin. He called for freedom from the landowners, the division of their estates, the restoration of the 'true monarch' and the 'old faith', the end to recruiting levies.

Pugachov's forces, amounting to between 20,000 and 30,000

men, waged guerilla campaigns on an unprecedentedly vast scale from Perm in the Urals to Tsaritsyn (now Volgograd) on the lower Volga. They took the important towns of Saratov and Kazan, but Ufa and Orenburg successfully defied the rebels' siege. Everywhere there was marauding and plundering, the burning of manors and hanging of landowners. Not until regular troops were released at the conclusion of the Russo-Turkish war did Catherine win back control of the insurgent areas. In 1775 the self-styled Peter III was brought in a cage to Moscow and his body dismembered. The whole episode exposed the hopelessness of Catherine's peasant policy.

THE GROWTH OF INDUSTRY

Economic development after Peter proceeded on traditional lines. But the tempo was quickened through the increase by about one-quarter in the size of the Empire, owing to annexations in the Crimea, Northern Caucasus, Ukraine, Poland, and Courland. The population increased by more than fifty per cent – from almost eighteen million in 1724 to some thirty million at the end of the century. This represented an increase in absolute numbers, not only the mere addition of the annexed populations. Of this total, serfs accounted for 9,900,000 males. Here was an absolute increase that largely resulted from the gifts of State lands made by Catherine and Paul to their respective favourites. The installation of a private owner had the effect of reducing State peasants to serfs. Very broadly speaking, a growing geographical differentiation took place in the functions of the serf. In the central provinces and to the north-west and south-west he was a performer of *obrok*, making payment in cash or kind to his owner. In the south, east, and north he was more likely working on a *barshchina* basis, toiling on his owner's land for so many days per week. This division of labour roughly corresponded to those areas where industry and agriculture were most developed.

The character of industry continued to be determined by cooperation with the State. The State was the market for the goods produced, it supplied capital and, to some extent, labour;

and it guaranteed certain monopolies by establishing high tariffs and sometimes by banning outright foreign products that would have competed with Russian products.

From the middle of the eighteenth century, the State introduced a number of positive measures to encourage industry. Under Peter, for example, the ports had been able to purchase grain for export only from certain specific areas, and internal trade had suffered from a number of customs barriers. But in 1753 these latter were abolished, and in 1762 free trade in agricultural produce was finally permitted. A number of monopolies were also abolished, and a general sentiment in favour of free trade characterized much of Catherine's economic policy: 'No affairs pertaining to commerce and factories', asserted a decree of 1767, 'can be conducted through compulsion, and a low cost of living is achieved as a result of a large number of sellers and an unhampered increase of goods.' A few years later another decree permitted 'everybody and anybody to start any type of mill and to produce in these mills any type of handiwork'.

All this does not signify Russia's sudden conversion to Manchesterian liberalism, but is rather related to the paramount interests of the nobility.

The fact is that some of the better-endowed nobility were beginning to go in for a type of manorial manufacture. And why not? There was cheap and abundant manpower available. In so far as the estate was a self-supporting unit, a certain degree of technical skill in handicrafts had been attained; raw materials, such as hemp, flax, wool, and hides, could be readily used. Equally important, no wages need be paid, since the serfs, in working on a manorial factory, were in fact merely performing their regular *barshchina*. In 1773, nobles owned sixty-six enterprises as compared with 328 merchant manufactures, that is, about twenty per cent of the total.

The freeing of the internal market also favoured two other types of low-level industry – the *kustar* or cottage industry, and the serf-owned enterprises. Textiles, including taffeta, silk, cotton, ribbons, as well as coarse military-type cloth, were the specialities of the first. But specialist goldsmiths, silversmiths, embossers, also produced luxury articles and leather goods to order. The forest areas would specialize in wooden objects, such as wheels and sleighs. In most cases, a merchant would make the rounds of the peasant homesteads and buy up goods for sale at town and village fairs.

The second category of small-scale manufacturers – the serf enterprises – sometimes grew out of the *kustar* type of production. A landowner, particularly if his estates were on *obrok*, had a powerful interest in securing the maximum output and productivity of his serfs so that they in their turn would be able to pay him increased *obrok*. As a result, the more enterprising serfs would be permitted and encouraged to set up small enterprises, employing the labour of other serfs or perhaps even hired labour. Sometimes it also became possible for a serf entrepreneur to buy his freedom from his owner. But to do this he would have to be exceedingly successful.

The paucity and inaccuracy of the statistics makes it impossible to show in detail the extent of industrial growth in the eighteenth century. But there is no doubt that it took place. Also important is the fact that new districts were opened up, primarily in the Urals. In 1766–7, out of 120 blast furnaces and iron-producing plants in Russia, seventy-six were to be found in the Urals. Most of these had been established in the second and third quarters of the century. The Urals were the predominant area for copper-smelting and iron and steel production.

On the other hand, the level of technique still remained very low. Water and horses were the main sources of power; charcoal was used exclusively in the smelting of pig-iron; and manual labour in the forging of iron. Whatever was achieved came about through the employment of a mass labour force, unskilled and miserably underpaid. This applied not only to the actual production, but also to such auxiliary operations as preparing and delivering the wood, water, and coal. Neither in industry nor in agriculture could a serf-economy be productive. As early as the eighteenth century, the arrested nature of Russian technology by comparison with that of Western Europe was abundantly clear.

POLAND AND TURKEY

The Great Northern War had made Russia into a great power, with interests extending from the Baltic and Germany in the north to the Caspian Sea in the south. In the course of its rise to this commanding position, Russia had eliminated Sweden and Livonia, made inroads into Poland, gained some influence in the North German states, and waged the first of many campaigns against Turkey. Each process was now to be taken a stage further. The expansion of Russia brought with it the gradual destruction of its neighbours to the west, north, and south.

After Peter's death, Ostermann, a shrewd and experienced Westphalian polyglot, controlled foreign affairs. In the 1730s German influence at St Petersburg enjoyed its heyday, and the Foreign Minister was one of its most distinguished representatives. He based his policy on an alliance with the Austrian Habsburgs in opposition to the French system of alliances, which embraced Sweden, Poland, and Turkey and functioned as a sort of *cordon sanitaire* against further Russian expansion.

Early on, Franco-Russian hostility became centred on the Polish question. Who would succeed Augustus II as King of Poland? A Russian or a French nominee? In the upshot, after a two-year struggle from 1733 to 1735, Russia succeeded in imposing Augustus III. As a by-product of the clash, Russian troops made their first appearance in Western Europe. They joined up with an Austrian army near Heidelberg. This war also saw the first clash of Russian and French arms, on the Polish battlefields.

The War of the Polish Succession brought with it a renewal of the Russo-Turkish struggle for Azov, the key to the delta of the Don, and to the whole river system of south Russia – the Dniester, Bug, Dnieper, Kuban. The four-year struggle from 1735 to 1739 partially reversed the unhappy verdict of Peter the Great's campaign of 1711. The Treaty of Belgrade of 1739 returned Azov and the adjacent territory to Russia, but the great fortress itself was destroyed. Taganrog was not to be fortified; and Russia, while acquiring the right to trade on the Sea of Azov, was obliged not to make use of its own ships but to allow its goods to be carried in Turkish bottoms. These results were all the less impressive by comparison with the number of Russian military successes during the war. But the defection of Austria from the campaign and its conclusion of a separate peace gave the Russians no choice. The war was also noteworthy in that it revealed that the Russian forces were far superior to the Austrian. This marked a significant stage in Russian military history.

In the Seven Years War (1756–63) Russia was one of the mainstays of the Austro-French coalition against Prussia and Great Britain. The fundamental Russian aim was to crush the Prussia of Frederick the Great. In the campaigns of 1759, 1760, and 1761, this was all but achieved, the Russians sweeping into East Prussia and securing some spectacular victories. They brought Frederick to the verge of suicide, notably after the battle of Kunersdorf in 1759. The following year saw Russian and Austrian troops in occupation of Berlin.

But just when Frederick's position was at its weakest, the fortuitous death of the Empress Elizabeth at the end of 1761 saved him from utter collapse. This brought to the throne Peter III, an idolater of Frederick the Great, in whose army Peter's father-in-law had served as a Field-Marshal. Within a few months, Russia had changed sides and taken up arms against its previous allies. No wonder Frederick the Great warned his successors in his Political Testament: *'après moi les souverains de la Prusse auront bien raison de cultiver l'amitié de ces barbares.'* The advice was well taken. From 1762 to 1914 there was no serious Russo-Prussian or Russo-German conflict. Russian withdrawal from the Seven Years War inaugurated a century and a half of peace between the two countries; it was of inestimable mutual benefit.

But what mattered more in the 1760s was that the conclusion of peace in the West gave Russia the necessary freedom of manoeuvre to move forward in Poland and Turkey.

This was primarily Catherine's inspiration. After having secured the assassination of her husband, Peter III, she took into her own hands the direction of foreign policy. 'I came to Russia a poor girl,' she declared. 'Russia has endowed me richly, but I have paid her back with Azov, the Crimea, and the Ukraine.'

The death in 1763 of Russia's earlier nominee, Augustus III, as King of Poland, very soon precipitated the Polish conflict once again. Once more, Russian bayonets united with the Polish landed aristocracy to ensure the 'election' of the Russian candidate – Stanislaw Poniatovski. Using as a front the protection of the rights of the Orthodox population, Catherine gained further influence in Poland, in 1768, by a Russo-Polish treaty which placed the Polish constitution under Russian protection. This was the prelude to the first partition of Poland in 1772, made in concert with Prussia and Austria. The Russian share took in the White Russian areas of Polotsk, Vitebsk, and Mogilev – altogether about 36,000 square miles and a population of about one and three-quarter millions.

In the meantime, Russia's growing influence in Poland was producing friction with Turkey. The whole territory north of the Black Sea, extending from the Danube to the Don and Caucasus, still lay under Turkish domination; and the Russians were beginning to menace this northern flank. Catherine also tried to undermine European Turkey from within by exploiting the Christian sympathies of the populations of Greece, Crete, Serbia, Montenegro, and Bosnia.

War broke out between Russia and Turkey in 1768. The war between the one-eyed and the blind, Frederick the Great uncharitably described it. In 1769 the Russians occupied Jassy and Bucharest and moved into the two principalities of Moldavia and Walachia. The next year, the Russian fleet, units of which were commanded by Admiral Elphinstone, formerly of the British Navy, sailed round Europe from the Baltic to the Aegean Sea. It totally destroyed the Turkish fleet off Chios. But these victories were not sufficient to bring the war to an end. Catherine constantly spurred on her armies to throw the Turks out of Europe altogether. But the war dragged on for another four years, its end eventually hastened by the Russian necessity to deal with Pugachov's uprising.

Legend:

▤ First partition of Poland 1772

▨ Second partition of Poland 1793

▧ Third partition of Poland 1795

⠿ Treaty of Kuchuk-Kainardji 1774, and the Treaty of Jassy 1792

0   100   200   300 Miles

4. RUSSIAN WESTWARD AND SOUTH-WESTWARD EXPANSION IN THE EIGHTEENTH CENTURY

The Treaty of Kuchuk-Kainardji of 1774 amply rewarded the Russian efforts by sea and land. The Crimea was wrested from Turkey, to become an independent state. Russia annexed Azov, Kertch, and Yenikale; acquired the right to establish consuls at will, to sail freely on the Black Sea and through the Bosphorus and the Dardanelles, and twisted certain provisions of the treaty into a basis for intervention on behalf of the Sultan's Orthodox subjects. The general effect of Kuchuk-Kainardji was to give Russia a firm hold on the northern shore of the Black Sea, and a foothold on the eastern shore, that is, on the northern Caucasus. In 1783 the logic of this was carried a stage further when the Russians annexed the Crimea; and in 1792 the Russian border advanced south-westwards to the coastal territory between the Dniester and the Bug. Here the great port of Odessa soon developed. Since the 1750s there had been some attempt to develop the new southern territories by settling foreigners in them. Serbs and Montenegrins were the first to come. A decree of 1762 invited all foreigners – again excluding Jews, as in Peter the Great's decree of 1702 – to install themselves on the empty lands of the Empire and acquire hereditary property on favourable terms. These immigrants came from Germany in the main, and by 1768 they had formed more than a hundred colonies, with a population of over 20,000 in the lower Volga areas near Saratov and Tsaritsyn.

These immense accretions of territory, coupled with the more settled conditions that prevailed, had one obvious political and commercial significance. They yielded the southern ports through which Russian grain could reach Western Europe, and England in particular. The war against Turkey was fundamentally a war for harbours. When the Industrial Revolution brought with it a vast increase in population, there were hopes in Russia that Eastern Europe would become the granary of Europe. The black-earth country – a wide belt of land stretching from a line Zhitomir-Kiev-Tula-Kazan in the north to the northern shores of the Black Sea and the Sea of Azov in the south – provided the raw material. Transport costs to the Baltic ports would have been prohibitive. The only practicable alternative was the Black Sea – and this was the Russian target. The export of Russian grain jumped from 32,000 chetvert [1] in 1717–18 to 400,000 chetvert in 1793–5. Most of it went through Riga and St Petersburg, but Taganrog was also growing. By 1793, one-fifth of Russia's cereal exports went through the Dardanelles.

Catherine's conquests also had a highly important demographic significance that did not become fully apparent until the nineteenth century. They provided the agricultural foundation for the threefold expansion of the Russian population in the nineteenth century. Up to the early eighteenth century the Russian State had developed in the main on a poor type of soil, the full exploitation of which was hampered by the inclement climate. Both these factors were particularly operative in the central Moscow area. The 300-per-cent population increase in the nineteenth century was largely dependent on the agricultural exploitation of the southern Russian areas annexed by Catherine. By the end of the eighteenth century, central and northern Russia had already become grain-importing areas, which lacked, furthermore, any of the mineral and industrial resources lying beyond the Urals. Had the newly annexed territories to the south not been available as grain-producing areas, agrarian over-population would have soon resulted. As it was, these areas provided the basis for the expansion of the population in the nineteenth century.

THE ENLIGHTENMENT IN RUSSIA

Voltaire wrote a history of Peter the Great and when he spoke of it to Madame de la Fontaine he said: '... *rien de plus ennuyeux pour une Parisienne que des détails de Russie.*' He would hardly have said this had he been speaking of the post-Petrine era. This was the period *par excellence* when three Empresses – Anne, Elizabeth, and, most of all, Catherine – set the tone of the social and intellectual life of Russia.

Their setting was St Petersburg. This had been the nominal capital since 1712, but only some two decades later did it become the real centre of Russian court life. The architectural

1. 1 chetvert = 5·8 bushels.

magnificence of the Winter Palace, the Hermitage, Tsarskoe Selo, the Taurida Palace, Gatchina, Peterhof, and the lesser but still opulent palaces of the wealthier, newly emancipated nobles gave a flamboyant and sumptuous air to the life of the capital. Not for nothing was St Petersburg known as the Palmyra of the north. The artists and architects who created this milieu were mainly foreign – Rastrelli, Quarenghi, Charles Cameron, Lamothe, Trombara – to mention but the best known. There was, of course, a seamy side to the magnificence, arising from the inclement climate and the damp atmosphere. An uncharitable French visitor recalled the phrase: 'It has been wittily enough said that ruins make themselves in other places, but that they were built at St Petersburg.'

The European Anne left little mark on the city. She divided her passions between those of the flesh, the shoot, and the hunt. In other respects her court followed the Petrine model which was itself a clear reflection of the High Renaissance, in its abundance of human monstrosities. Midgets, giants, hunchbacks, cripples, the deformed in body and mind, were a ghastly feature of Anne's entourage.

Elizabeth was a more sympathetic character. Sir George Macartney, who came to Russia as British Envoy Extraordinary shortly after her death in 1761, rather unkindly described her as 'abandoning herself to every excess of intemperance and lubricity'. In actual fact, pleasure and piety dominated her life. To Elizabeth the cares of State were non-existent. Her days she devoted to relaxing *en déshabille*; her nights to love-making and dancing and masquerades. She had 'not an ounce of nun's flesh about her', said Finch. A particular *penchant* was to have the soles of her feet tickled. The minuet was the most popular measure, as everywhere in Europe. At St Petersburg it shared pride of place with the quadrille (often masked). At some of the court balls, Elizabeth combined the fetish of the foot with transvestist tastes; the men must dress as women and the women as men. In this way Elizabeth could display her tiny feet and shapely legs which would otherwise have been concealed beneath the enormous hooped skirts of the period. Her favourite male costume was that of a Dutch sailor. This was a tribute to the memory of Elizabeth's father, Peter the Great. The young Catherine, the future Empress, has described in her memoirs something of the strange sights produced by these entertainments:

The women looked like scrubby little boys, while the more aged had thick short legs which were anything but attractive. The only woman who looked well and completely a man was the Empress herself. As she was tall and powerful, male attire suited her. She had the handsomest leg I have ever seen on any man and her feet were admirably proportioned.[1]

Catherine, in her turn, took no less enjoyment in the pleasures of the flesh than did Elizabeth. The older she got, the younger her lovers. All together, the total ran into fifty-odd. But what really made Catherine different from Elizabeth was her compelling interest in State affairs. Never did she allow private concerns to clash with public. Like Frederick the Great, or Peter, or Joseph II, or Maria Theresa, Catherine saw herself as an enlightened despot of the eighteenth century, sacrificing her own well-being to her people's. By intellect and taste she would have been perfectly at home with Voltaire at Ferney, with Gibbon at Lausanne, with Walpole at Strawberry Hill. Her regular correspondents included such representative men of the age as Voltaire, Diderot, D'Alembert. Her favoured reading was Montesquieu's *L'Esprit des lois* and Beccaria's *Crime and Punishment*.

Chance had it, however, that Catherine was no mere member of the cosmopolitan European intelligentsia of the eighteenth century, but also the Empress of Russia. And this dual role cruelly revealed the discrepancy that distinguished her practice from her professions. 'Let a man criminate himself,' said Lord Acton. Catherine indeed did so. No other ruler, perhaps, has been so thoroughly discredited by his or her own writings and decrees. This was all the more fatal in that Catherine looked on herself as Russia's prime propagandist and was for ever feeding her distinguished correspondents glowing accounts of the

1. *The Memoirs of Catherine the Great*, English translation, London, 1955, pp. 185–6.

country – 'no single peasant in Russia could not eat chicken whenever he pleased,' she assured Voltaire, just a few years before Pugachov's rebellion revealed the desperate mood of large sections of the population.

But with all its obvious opportunities for satire, Catherine's long reign marked a definite stage in the education of the top stratum of the Russian nobility. French became the language of the court. With the language went the thought of the Enlightenment. 'The first lessons in French that Russians learned were lessons in blasphemy,' said Joseph de Maistre of the all-pervasive Voltairian influence in Russia. French tutors were very much *à la mode* in the families of the wealthier *dvoryane*. La Harpe taught Catherine's own grandson. In the Saltykov family the tutor was Charles-François Masson, who actually attributed Russian Gallomania to the prevalence of French tutors: '*Les Russes, presque tous élevés par des Français, contractent, dès leur enfance, une prédilection marquée pour cette nation.*' At the same time, the freeing of the nobility from State service coupled with the abolition of restrictions on foreign travel led to the growth of a sizeable aristocratic Russian colony in Paris. Count Paul Stroganov, for example, was the archivist of the Jacobin Club. The Golitzin, Shuvalov, and Vorontzov families were as well known in Paris as in St Petersburg. Gambling and libertinage distinguished them in both capitals.

The eighteenth century did little for Russian culture *tel quel*. The first Russian theatre, *maître-de-ballet*, actress, were little to set against the tedious and perfunctory official literature of the period, or the plethora of French and Italian opera and ballet companies. What did emerge in the eighteenth century, however – and this had supreme importance as the pre-condition and essential instrument of the literary efflorescence of the nineteenth – was a Russian literary language. The remarkable genius of Michael Lomonosov largely brought this about. The son of a White Sea fisherman, he developed polymathic gifts as professor of chemistry, scientist, philologist, and historian.

In Peter's time a certain linguistic confusion had developed through the decay of Church Slavonic and its alienation from popular understanding, and also through the introduction of secular literature from the Ukraine and the West. The suicide of one of the translators appointed by Peter the Great to render into Russian a French work on horticulture illuminated the linguistic crisis. He was simply unable to find in Russian the equivalents of the French terms. Lomonosov broke down this isolation by amalgamating the literary language of Church Slavonic with the non-literary of spoken Russian. A virtually new language was created.

But if the eighteenth century saw little development in Russian culture, it was very different where the flow of ideas was concerned. If Voltaire's works could circulate, and the *Encyclopedia*, and Blackstone's *Commentaries*, and *L'Esprit des lois*, if Russians could study abroad – then it could be only a question of time before some attempt would be made to apply to the Russian reality some of the ideas garnered in Leipzig, Paris, Geneva, Dresden.

The French Revolution, for example, was enthusiastically welcomed by many people in St Petersburg. Ségur, the French Minister, saw passers-by in the streets of the capital excitedly embracing each other when news came of the fall of the Bastille. He himself was greeted and congratulated, while walking along the Nevsky Prospect, by Catherine's grandsons, Alexander (the future Emperor) and Constantine, his brother.

Catherine herself now saw the danger she was running, and tried to ward it off. She broke off diplomatic relations with France; she put all French-speaking foreigners under police supervision; stopped the sale of the *Encyclopedia* and confiscated a new Russian translation of Voltaire's works. Frenchmen were anathema. To Grimm she wrote: '*Il faudra faire jeter au feu tous leurs meilleurs auteurs, et tout ce qui a répandu leur langue en Europe, car tout cela dépose contre l'abominable grabuge qu'ils font.*'

But by now the damage was done – not so much by actual revolution, but by the slowly gathering impact of French liberal ideas. In 1790, through some aberration of the censorship, about fifty copies of a work entitled *Journey from St Petersburg to Moscow* were put on sale. The author was a certain Alexander Radischchev, assistant-director of the St Petersburg Customs

House, and a member of a family of provincial serf-owning nobility. After serving in the Corps of Pages, he had been sent by Catherine to study at Leipzig University. Here he read for preference the works of Rousseau, Mably, and Helvétius.

The *Journey*, printed on a private press, permission for which had only recently been given, is a strange, discursive hotchpotch of prose and verse, autobiography, and descriptive narrative. The technique is ostensibly taken from Sterne's *Sentimental Journey*. But what a difference between the journey through France and Italy and the journey from St Petersburg to Moscow! Radischchev was no revolutionary, rather a sentimental Utopian, who looked 'across a century', as he himself put it. But this did not prevent him from penning the most outspoken criticism of Russian life that had yet appeared.

What did the traveller find on his grisly journey from the new to the old capital? He found serfs forced to toil six days a week on their master's land. He found officials and judges who would sell a verdict or a decision. He found corrupt businessmen. He witnessed a serf auction, where human beings were knocked down like cattle to the highest bidder. He saw a round-up of serfs, press-ganged into army service, as in eighteenth-century England. And everywhere he was overcome by the tragic results of autocratic rule.

Radischchev was sentenced to death at Catherine's instigation, but she later commuted his punishment to ten years' exile in central Siberia. He was, in fact, released after seven years, on the accession of the Emperor Paul.

This exile epitomized the breach between the autocracy and the educated classes. By the same token, Radischchev's fate signalized, as Berdyayev has truly said, the birth of the Russian intelligentsia: 'I looked around me – my soul was afflicted by the sufferings of humanity ... I felt that all may take part in helping their fellow-men.' These words, from Radischchev's dedication, became the leitmotive of generation after generation of the intelligentsia.

CHAPTER 9

The Reign of Alexander

ONE fine autumn day – it was either Sunday or some festival – we had been at Mass in our parish church dedicated to the Assumption of the Virgin, and were returning home. We had just mounted the high step outside our house, when suddenly we noticed some commotion and loud talking among the people returning from church. Next, along the street at full gallop came a Cossack orderly; to all whom he had passed he cried out, 'Go back to the church to take the oath of allegiance to the new Emperor!'... The cathedral bells began to ring, the bells of ten other churches chimed in, and the sound spread all over the town.[1]

In this undramatic fashion the news of Catherine's death reached one small Russian town – Ufa. It lay in the heart of the Bashkir country on the southern Urals; it is now a centre of oil refineries and motor manufacture. The new Emperor to whom the population had to swear allegiance was Paul, Catherine's son. He did not last long. He betrayed symptoms of severe mental unbalance; and he attempted to clip the powers of the nobles by making the throne hereditary in the Romanov family and by limiting the nobles' right to withdraw at will from Government service. Their estates were subjected to tax, their legal immunity from flogging removed, and their military service made less nominal. A Guards commission was no longer a sinecure.

In 1801 a group of conspirators entered Paul's bedchamber and, with the knowledge of the Grand Duke Alexander his son and heir, murdered the Emperor.

The nocturnal *coup* marked, in a highly symbolical way, the end of a phase in Russia's political evolution. Thereafter assassinations were to take place in the open streets and would cease to be the prerogatives of palace cliques and Guards officers.

1. Sergei Aksakoff, *Years of Childhood*, English translation, Oxford, 1923, p. 158.

This widened concern with politics was one important symptom of the new age. In foreign policy it saw the growth of Russia from a European power into a world power; culturally, it saw an unprecedented flowering of thought and literature; in economic life, there was a significant increase in industrialization; politically, it saw the growth of a revolutionary movement that at first embraced a few scattered individuals, but later coalesced into highly organized parties. All these developments had one factor in common : they all called into question the continued existence of the autocracy in the form in which it had hitherto existed.

One of the first problems to confront the new Emperor, Alexander I, was Russia's relationship to Napoleon. From the first, of course, the Russian Government had been hostile to the French Revolution and its military campaigns. But Catherine had taken no part in its suppression, preferring to occupy herself with the Second and Third Partitions of Poland (1793 and 1795). Paul was more active, and in 1799 Marshal Suvorov's campaign in Northern Italy helped to throw the French back. But his later defeat in Switzerland, coupled with Russo-British friction in Holland and in the Mediterranean, where the Russian fleet was operating, led to Russian withdrawal from the campaign. Moreover, Napoleon's overthrow of the Directory and his installation as First Consul helped to convince Paul that the dangerous, explosive force of the Revolution had been tamed. At home, the war was highly unpopular among the nobles, who lost, in England, their chief foreign market for Russia's exports of timber, flax, and hemp.

Soon after his accession, Alexander made peace with Napoleon and also with England. But this came to an end when Russia entered Pitt's Third Coalition in 1805. Two years later, first Prussia and then Russia were defeated by Napoleon. Yet the defeat was not so overwhelming that Alexander could not negotiate peace terms with Napoleon; and this was done at Tilsit in 1807, on a raft moored in the River Niemen. Russia, indeed, lost no territories. Russo-French peace, and also an alliance, were sealed largely at the expense of Prussia, which lost all its Polish acquisitions. These were formed into the

Duchy of Warsaw. Russia annexed the Bielostok district. On the other hand, it also had to enter Napoleon's continental system and to suffer all the inevitable consequences to its export trade. Partial compensation came in Russia's enjoyment of a free hand in the north-west and south-west.

In the one, Finland was seized from Sweden, and in the other Bessarabia from Turkey. Farther to the east, Paul's annexation of Georgia in 1801 was followed by a war with Persia in the Caucasus. Here too there were Christians to protect; here also lay a land-route to India and the East. After a long, swaying, wayward struggle in this wild mountainous country, Persia at length acknowledged Russian sovereignty over a large area stretching from the Black Sea to the Caspian. In later years this was shown to be the oil-land of the Caucasus.

In the West, meanwhile, the alliance with France was breaking down – mainly as a result of Napoleon's aggrandizement of the Duchy of Warsaw. In June 1812, the Grand Army crossed the Niemen into Russian territory. And then the truth was proved of a remark made to Alexander by Rostopchin, the Governor of Moscow: 'The Emperor of Russia will always be formidable in Moscow, terrible in Kazan, and invincible in Tobolsk.'

But it was not only Russian geography and climate that defeated Napoleon and the Grand Army. The invasion of 1812 created one of those occasions, rare in any country's history, when a genuine movement of national resistance emerged to throw back the invader. It was something akin to the rally that had thrown back the Poles in 1612; or to the similar national upheaval that Napoleon had roused in Spain. Nobles and merchants hastened to raise and finance regiments; guerilla bands spontaneously emerged to harry and harass Napoleon's flanks. The supreme Russian sacrifice came with the burning of Moscow. (It is uncertain whether this took place by design or chance.) The Emperor of Russia had no need to withdraw to Kazan or Tobolsk. The flames of burning Moscow consumed Napoleon's hopes of a swift conquest. He had no choice but to retreat.

The Russian advance did not stop at throwing Napoleon back

over the Niemen. It carried on into Western Europe, and at the Congress of Vienna in 1815 secured for Alexander a powerful voice in the settlement of Europe. A certain *folie de grandeur* now took possession of his soul. Through the Holy Alliance with Prussia and Austria, later joined by most of the European States, he dedicated himself to the duty of introducing into diplomacy 'holy faith, love, justice and peace', as a criterion of interstate relations.

This notion had little importance, and certainly did not affect Alexander's attempts to secure for Russia the whole of Poland as an autonomous kingdom under Alexander's protectorate. This aroused the opposition of England and Austria. In the end Alexander had to moderate his claims, and to share Poland with Austria and Prussia. But he retained the major portion. The frontiers in Europe with which Russia emerged from the Napoleonic Wars remained unchanged until 1914, apart from some readjustments in Bessarabia.

THE DECEMBRIST CONSPIRACY

Alexander came to the throne a reputed liberal, and in his early years he went some way to justify this reputation. He abolished the security police; he removed the ban on foreign travel and permitted the importation of foreign publications. He even dallied with constitutional reform, and charged Michael Speransky, the son of a village priest who developed into a brilliant administrator, to draw up a scheme for the improved government of the Empire. This was ready in 1809 and, in essence, aimed at reconciling the autocracy with a system of law by introducing the separation of legislative, executive, and judicial powers. All these would flow from the prerogatives of the autocracy, but the latter would to some small extent be limited by a State Duma elected on a property franchise, totally excluding the serfs. In the end, all this was watered down into a State Council appointed by the Crown and equipped only with advisory powers. It was never any curb on the autocracy. Of Speransky's other proposals, the most important to be adopted was a reorganization of the executive functions of the ministries; this amounted to a relative modernization of the Imperial bureaucracy. As before, despite the bureaucracy's immense importance as a centripetal force, it remained corrupt, irresponsible, and inefficient. It enjoyed none of the prestige attaching to the Prussian or British Civil Service. Hostility to the bureaucracy was further stimulated by the presence in its ranks of a high proportion of Germans.

Alexander's treatment of the serf problem was on a par with his treatment of Speransky's project. It was, of course, an infinitely more delicate topic than any constitutional matter. 'The Emperor cannot rule without slavery,' prophetically wrote de Maistre (meaning by 'slavery' serfdom). To touch serfdom was to touch the very foundation of the State. Yet a number of factors, gathering strength from the beginning of the century, were making it less and less possible to avoid dealing with it by some means or other.

Was serfdom an economic proposition? The Free Economic Society of St Petersburg which held a competition for essays on this subject in 1812, found that the authors of the fourteen manuscripts submitted were equally divided on this point. The conclusions of the winners of the first and second prizes, both of whom supported free labour, were emphasized by the fact that the total agricultural yield remained virtually stationary between 1810 and 1870. Indeed, there was some attempt, particularly in the south-west, to introduce capitalist farming technique with livestock rearing, the use of modern machinery, and the employment of hired labour. But all this remained a drop in the ocean as compared with the vast extent of grain farming. The basic requirements of capital, and a free, skilled, and productive supply of labour were both lacking.

The precarious economic position of the landowners also militated against agricultural productivity and progress. By 1843, for example, more than fifty-four per cent of noble-owned lands were mortgaged to different State institutions. Many estates were also encumbered by private debts.

Serfdom not only posed economic problems, but also had immense political and moral repercussions. However deep-rooted and old-established, it was none the less acknowledged to be an evil that demanded some attention. What was the secret of emancipation without tears? This inescapable and tortuous conundrum dominated all thinking on the question in the first half of the century and formed the leitmotive of the official and unofficial committees that Alexander established. In general, these committees were quite unproductive. But they did at least serve to illuminate the complexity of the problem and to reveal the bitter opposition that any proposed solution would inevitably provoke. It was only in the fringe areas of the Empire – Courland, Livonia, Estonia, and in Poland – that serfdom could be abolished, or its effects mitigated by a system of 'inventories' that codified more precisely the respective obligations of landlord and peasant.

All that was done in Russia itself was to enact a number of minor measures that did to some extent ameliorate the position of the serfs. But they only tinkered with the problem. In 1801 a decree made it illegal to advertise serfs for sale. Two years later an interesting attempt was made to create a class of free agriculturists whereby a landowner wishing to emancipate his serfs was obliged to provide appropriate areas of land, on agreed terms, for the use of the freed serfs. But by 1855 no more than one and a half per cent of the serf population had been freed in this way. After 1809, landowners could no longer sentence their serfs to penal servitude.

As Alexander grew more and more engrossed in the campaign against Napoleon, as he fell more and more deeply under pietistic and mystical influences, as his self-appointed mission of European saviour absorbed him more and more passionately, his liberalism dwindled into a mere concern for the *status quo*. This was the equivalent at home of the Holy Alliance abroad. To all appearances, there was little to distinguish the state of Russia at the beginning and at the end of Alexander's reign.

But such an appearance was wholly and utterly deceptive. The fact is that Alexander's campaign against Napoleon had unintentionally given decisive impetus to a new generation of thinking Russians. These were the sons of those of the eighteenth century who had read the *philosophes* and drunk of the

waters of Enlightenment. They were Guards officers, land-owning nobles, senior officials, and the offspring of the wealthier
class of merchants and professional men. But where the fathers
had become critics through their reading, the sons had seen the
West at first hand. Radischchev was the first of the 'repentant
noblemen'; after the Napoleonic Wars he had a host of progeny.
Not for nothing did Herzen say that only in 1812 did the
veritable history of Russia begin.

The Napoleonic Wars and the Russian entry into Paris
brought large numbers of the educated nobility for the first
time into immediate contact with unfamiliar conditions of a
nature infallibly destined to arouse Russian self-doubt and self-
criticism. 'The campaigns of 1812–14,' wrote Prince Volkonsky
in his memoirs – and this assertion is typical of many – 'brought
Europe nearer to us, made us familiar with its forms of state,
its public institutions, the rights of its people. By contrast with
our own state life, the laughably limited rights which our people
possessed, the despotism of our régime first became truly
present in our heart and understanding.' If even Poland and
Finland had constitutions, why should not Russia also? Two
other factors deepened the feeling of dissatisfaction: first, the
further contrast between Russia's deplorable governmental system and her role at the Congress of Vienna as one of the great
powers of Europe; and secondly, the sentiment of popular
unity, generated by the national upsurge of 1812, which led in
turn to the notion of a duty owed by the higher orders to the
lower, and to the serfs in particular. Rostopchin, the Governor
of Moscow in the 1820s, dryly commented on this phenomenon
from the point of view of the régime: *'Ordinairement ce sont
les cordonniers qui font des révolutions pour devenir grands
seigneurs; mais chez nous ce sont les grands seigneurs qui ont
voulu devenir cordonniers.'*

The first fruits of the revolutionary spirit of the *grands seigneurs* was the Decembrist revolt, so called because it took place
in December 1825. It was a Janus-like conspiracy, a palace *coup
d'état* of the eighteenth century compounded with the revolutionary theory of the nineteenth. It looked both to the past and
to the future. It fell between two worlds, a dichotomy that may

well be the ultimate reason for the failure of the attempted revolt.

The antecedents of the Decembrist movement reach back about a decade, to the immediate post-Napoleonic period. In 1816–17 a number of highly-placed officers, the scions of aristocratic families – the brothers Muravyov, Colonel Pestel, and Prince Serge Trubetzkoy – formed a Union of Salvation in St Petersburg. Its objects were to abolish serfdom and to install a constitutional régime. Before long the Union of Salvation died. In doing so, it gave birth to the Union of the Public Weal. Here again, military men predominated in a membership that fluctuated between fifty and 200. The programme, although it was never homogeneous and fully accepted, showed a certain move leftwards. Certain members, led by Pestel, saw no solution for their problems save in the forcible establishment of a republic.

Again the Union broke up after a few years, to be reformed in St Petersburg at the end of 1822 under the name of the Northern Society. In the meantime, a Southern Society had also come into existence under Pestel, who was stationed with his regiment in the Ukraine. In 1825 the Southern Society merged with the equally conspiratorial Society of United Slavs, composed of delegates from the Slav peoples of the west and south – Czechs, Slovaks, Croats, Slovenes, Serbs, and Ruthenians. The Poles were affiliated through the Polish Society of Patriots.

Pestel, the son of the German-born Governor of Siberia, a former student of Dresden University, knowledgeable in French revolutionary theory and practice, and a veteran of the Borodino battle and other Napoleonic campaigns, was by far the most interesting Decembrist. He held that the Imperial family must be assassinated and the autocracy replaced by a republican system based on universal suffrage. Pestel, known to his critics among the Decembrists as 'a Bonaparte, not a Washington', was clearly a Jacobin type. His *Russkaya Pravda* – Russian Truth, or Justice, or Law – which propounded the ideal programme for the morrow of the successful revolt, envisaged a régime that combined compulsory republican virtue with centralized government. He demanded the abolition of serfdom and the division of the land into publicly and privately owned sectors, with every family having enough land to be self-sufficient. Pestel also urged the need for free enterprise, equality before the law, the jury system, equal military service for all classes, and the abolition of a standing army. By means of a police force and secret spying organization, Pestel's government would forbid all private associations of citizens. The authoritarian form of State was further expressed in a policy of Russification extending to all the peoples of the Russian Empire, except the Poles. The Jews too, Pestel argued, must assimilate or undergo forced emigration to Asia Minor. There was a converted Jew among the Decembrists of the Northern Society, a certain Gregory Peretz, and he may have suggested this to Pestel.

By comparison with this authoritarian scheme, Nikita Muravyov, the theorist of the Northern Society, son of a senator and heir to great estates, propounded the idea of a federated, constitutional monarchy on the pattern of the thirteen original States of North America, each a self-governing entity. Only four ministries would, he argued, represent the proposed Russian federation as a whole – the Ministries of Foreign Affairs, Finance, War, and the Navy. Civil liberties, trial by jury, and the freeing of the serfs, though not necessarily with grants of land, were other features of Muravyov's programme. Though notably less radical than Pestel, Muravyov was not the less a revolutionary in the Russian context.

Secret propaganda, preparation, and recruitment filled the years 1823–5. Officers only were enlisted; the soldiers would obediently follow their superiors. But in which direction would their superiors lead them? After much debate a plan was finally concocted, timed to come into operation in May 1826 when the Tsar was expected to attend a troop review in the south. This would provide the occasion for his assassination. The arrest of the Grand Dukes, the occupation of Kiev, a march on Moscow, and the assumption of control over the Guards regiments in St Petersburg – all this would follow fast.

But no troop review in the south took place. Alexander inconveniently died at Taganrog on 19 November 1825 (old style). A week later the news reached Moscow. A minor dynastic crisis followed. It lasted about three weeks. Constantine, Alexander's elder brother, who was apparently the legitimate heir, had forfeited his claim to the throne by concluding a morganatic marriage. The Decembrists hoped to exploit the confused interregnum to pull off their own *coup* before Nicholas, the dead Tsar's younger brother, ascended the throne.

The hurriedly revised plan called for a revolt on the day when the soldiers were to take their oath of allegiance to Nicholas. But there was treachery. Field-Marshal Baron Ivan Diebitsch knew of the plot from a non-commissioned officer named John (Ivan) Sherwood, who was of English origin. Information came from Jacob Rostovstev also, an officer whom Prince Eugène Obolensky had unsuccessfully tried to enlist in the Northern Society. 'In the early hours of the day after tomorrow' (that is, the day appointed for the taking of the oath), Nicholas wrote to Diebitsch, 'I shall be either a sovereign or a corpse.'

In actual fact, he became the sovereign. Treachery was not of itself fatal to the plot. What did kill it at the roots was the collapse of the plotters' morale. Some defected at the last minute; others, after leading their troops on to the Senate Square in St Petersburg, where the ceremony of allegiance was due to take place, failed to give the order to rise. For about six hours, until the early winter dusk drew in, loyal and insurgent troops faced each other. Hardly a salvo came from either side. At one point, Nicholas ordered carriages to be ready to take the Imperial family to safety. But he never lost his head; and after dusk had fallen and after the mutineers had rejected the offer of capitulation, he had cannon trained on them, and in two hours had cleared the Square. In the south the march on Kiev came similarly to grief.

The authorities interrogated and investigated about 600 people, many of whom confessed, gave way to mutual recrimination, and implicated their fellow conspirators. Nicholas himself displayed rare skill in eliciting from the prisoners the maximum information. Pestel was one of the few to maintain his sang-froid. 'We wanted to harvest before we had sown,' he told his judges. In the end five people were sentenced to death by hanging and more than a hundred exiled to Siberia or to army service in the ranks.

This marked the official end of what Nicholas called 'a horrible and extraordinary plot'. No mention of it was allowed to appear in the Press, and everything was done to destroy the very memory of the abortive revolt. But it survived as a myth to inspire all future rebels against the régime – the intelligentsia of the forties, the Nihilists of the sixties, the Populists and Anarchists of the seventies, and the Marxists of the eighties. The Resident British Minister in St Petersburg sent Canning a remarkable diagnosis of failure and an equally remarkable prophecy:

The late conspiracy failed for want of management, and want of head to direct it, and was too premature to answer any good purpose, but I think the seeds are sown which one day will produce important consequences.

CHAPTER 10

The Reign of Nicholas

To the new Tsar, twenty-nine-year-old Nicholas, the Decembrist revolt came as a traumatic shock. From the first, he was resolved to preserve the *status quo* intact. Not that Nicholas was not *au courant* with all the abuses, injustices, and corruption stigmatized by the Decembrists. On the contrary, he took the greatest pains to familiarize himself with their criticisms. But he did very little about them. It was this Russian government that de Custine compared to 'the discipline of the camp – it is a state of siege become the normal state of society'.

All initiative at any level but the topmost was stifled through the operation of the censorship and the closest possible supervision of the activities of the populace. The revolts of 1830–1 in Belgium, France, and Poland, and the revolutions of 1848–9 throughout the Continent, gave Nicholas renewed and reinforced incentive to preserve Russia inviolate. The slogan – 'Orthodoxy, Autocracy, Nationality'–coined by Count Uvarov, the Minister of Education, was intended to serve as an ideological dam that would hold back all critics of the existing order. In one way or another a whole generation of Russian thinkers and writers suffered from this oppressive régime: Pushkin, Lermontov, Herzen, Belinsky, Turgenev, Bakunin, Dostoevsky – these were but a few of the most prominent. The autocracy had virtually no supporters of any distinction among the intelligentsia.

The characteristic institution of Nicholas's reign was the Personal Chancellery of His Imperial Majesty. It consisted of a varying number of sections, of which the Third by far exceeded all the others in importance. The very formation of this section, concerned with internal security, testified to Nicholas's distrust of the regular governmental machinery, and also to the embarrassment felt by that lonely man, in his austere room in the vast Winter Palace, at not knowing what was going on among his subjects. Where no free expression of public opinion was permitted, the autocrat lived in a vacuum, beset by feelings of uncertainty and insecurity. The Third Section, it was hoped, would overcome the gap that divided the autocrat from the knowledge of what was being felt and thought by his subjects. Most important of all, it would keep track of any subversive ideas and institutions. This was the true and innermost *raison d'être* of the Third Section. Foreigners, dissenters, criminals, sectarians, writers – any exponents of ideas would come under its supervision. The Section was headed by General Benckendorff, who had early convinced Nicholas that the Decembrist conspiracy proved the inadequacy of the existing police force. This was an obvious challenge to create improved forces.

These were of two kinds: first, a uniformed gendarmery covering the whole empire, organized and officered on military lines in five (and then, eight) regions; second, the Section operated through a corpus of secret informers. Very often the Section functioned not only as a police force proper, but also as a judicial body with the power to punish or to exile.

But for all the censorship, the Third Section, the supervision of universities and schools, the prohibition of foreign travel, the propagation of official patriotism, there were three main disruptive forces at work against which all the Government's measures were helpless. These were the serf problem, the development of industry, and the growth of the intelligentsia.

Nicholas did, it is true, introduce certain palliatives in the treatment of the serfs. He prohibited the sale of serfs without land as a means to settling private debts, and also auction-sales of serfs that would have dispersed families. Serfs whose owners went bankrupt received the right to buy their liberty. On the State lands, Kisselev, the Minister of State Domains, carried through certain reforms which equalized the peasants' land allotments and created some sort of welfare system – schools, a medical and fire service, and agricultural advisory centres.

None the less, force and violence were becoming more and more a factor in the development of the peasant question. Certainly there was nothing remotely comparable to the Pugachov revolt, but there was a significant uninterrupted rise in peasant rebellions during the reign. Between 1826 and 1854, 712 peasant outbreaks were officially reported: from 1826 to 1834, 148 outbreaks; 1835–44, 216; 1845–54, 348. The outbreaks would run the whole gamut of violence, from illegal timber-cutting to the destruction of crops and the murder of land-owners. Violence was at its worst in the old central districts – Vladimir, Ryazan, Tambov, Voronezh, Penza, Saratov, and in the Urals. These were the oldest homes of serfdom and of forced labour.

The first intimations of the industrial revolution came to Russia at the same time as serfdom was becoming increasingly vulnerable to upheaval. In the first half of the century, industry, and particularly light industry, developed at a much faster rate than agriculture. Production of pig-iron barely doubled (as compared with a thirtyfold increase in Great Britain over the same period.) There was almost no coal, fuel, or petroleum industry and very little in the way of non-ferrous metallurgy. This signified a tremendous reversal of the trend of the previous century, when, for example, Russian production of cast iron had been at least equal to that of England. The reason for this change of fortune seems to have been the unfavourable loca-tion of coking coal in the Donetz Basin, far from the Ural iron-ore deposits.[1] It was certainly no shortage of manpower that retarded Russian economic development.

In fact, one of the clearest indices of the growth of industry was the growth in the number of workers, both on a posses-sional and hired labour basis. The number of manufactures rose from 2,402 in 1804 to about 14,000 in the second half of the 1850s. The number of workmen increased correspondingly. In 1804 there were 225,000; in 1825, 340,000; in 1860, 860,000. The proportion of hired labour in these totals rose respectively from twenty-seven and a half per cent to thirty-three per cent, and then to sixty per cent. The accelerating tempo of indus-trialization is shown in the rate of growth in hired labour. It increased in absolute figures 1·9 times from 1804 to 1825, but

1. See A. Baykov, 'The Economic Development of Russia', *Economic History Review*, 2nd Series, Vol. VII, No. 2.

4·6 times from 1825 to 1860. Over the whole period 1804–60, there was almost an 8·6 increase in the number of hired labourers.

Two industries stood out for their rapid growth – cotton textiles and beet sugar. The former was distinguished for its use, from 1840 onwards, of hired labour exclusively. It also benefited greatly from the impact of the most modern English machinery (after 1842, that is, when the English ban on the export of textile machinery was lifted). Knopp, an English entrepreneur, built and maintained the majority of Russian cotton factories. The most developed industrial centres were Moscow, Vladimir, and St Petersburg.

THE INTELLIGENTSIA

'One word of truth hurled into Russia is like a spark landing in a keg of gunpowder,' said de Custine. Those who hurled this spark of truth were the intelligentsia. At this time it was still very largely confined to the aristocracy, to the same class that had produced the Decembrists. But the latter's defeat in 1825, combined with Nicholas's repressive policy, brought with it a general withdrawal from the political struggle and a shift of interest from politics to history, or rather, to the meaning of Russian history with its manifold political implications.

This was the great age dominated by the historical theories of Hegel and Schelling. As Lord Acton pointed out, the social function of a historical philosophy such as that of Hegel in Germany, of Burke in England, of de Maistre in France, was to form the 'resistance to the revolutionary rationalism' of the eighteenth century, to decry the power of the unaided human mind in its capacity to influence the development of society, and to exalt in its stead the essentially irrational and unthinking weight of the past and of tradition, against which it was futile for the individual to pit himself.

But what was a reactionary philosophy in the West had, characteristically, revolutionary connotations in Russia; at the level of ideas, that is to say, and not in terms of practical politics. For what was the Russian past, the Russian tradition, in which men were to seek a guide to the present and future? Here lay the core of the tense debates in the thirties and forties between Slavophiles and Westerners. The conflict was no new one in Russian intellectual history. What Herzen called 'wounded national feeling, dark recollection, healthy instinct' had characterized a certain movement of opinion and cast of thought from at least the seventeenth century. The Old Believers had been Slavophile, as against Nikon, the Westerner; so had the opposition to Peter the Great; so had a certain Prince Schcherbatov in his opposition to the Gallomania of the eighteenth century. All such critics had manifested what is known to sociologists as 'Utopianism of the past' – a tendency to hark back to the good old days. Precisely the same conflict would run through the left-wing and socialist movement of the latter part of the nineteenth century.

What made the conflict particularly acute at the beginning of the century was the nearness of the West and the impact of Western constitutional ideas and ideals. The first fruits of this were the Decembrists, Westerners to a man. To them it was axiomatic that the light came from the West, whether France, England, or the United States.

But now the conflict entered a new phase. Hegel himself had slighted the Slav world as 'an intermediate essence between European and Asiatic spirit'. This limbo left Hegel's students all the more freedom to construct their own version of Russian history. The issue was clearly posed: did Russia belong to Western society, albeit in a retarded state; or did Russia embody a civilization *sui generis* that must be saved from the contamination of Western influence?

Peter Tchaadayev, an ex-officer in the Hussars and friend of the Decembrists, a dandy, an *habitué* of the aristocratic salons of St Petersburg and Moscow, brought the conflict to the fore with *A Philosophic Letter*. It was written in 1829, but not published in the Moscow journal *Telescope* until 1836. The 'letter' provoked an immense furore – 'what an alarm in the salons', exclaimed one contemporary. Tchaadayev was officially declared insane and subjected to a sort of house arrest. What had he done? He had simply condemned Russian history *en bloc* as sterile and worthless because of its separation from Western influence. Russia had remained apart from the world's developments, stagnant in its isolation.

Confined in our schism, nothing of what was happening in Europe reached us. We stood apart from the world's great venture [Tchaadayev argued]. . . . While the whole world was building anew, we created nothing: we remained crouched in our hovels of log and thatch. In a word, we had no part in the new destinies of mankind. We were Christians, but the fruits of Christianity were not for us.

Tchaadayev's remedy was a *rapprochement* with Catholicism, the medium whereby Russia should rejoin the West.

Here was a clarion call to the Westerners. Their members included Herzen, the brilliant publicist, Belinsky, the founder of Russian literary criticism, Turgenev, the novelist, Granovsky the historian, and Bakunin, the future anarchist. Whatever their other differences, there was a fundamental belief in the urgent necessity for closer contact with the West, where the virtues of free thought, rationalism, individual liberty, the values of science existed, and could serve as the means and the model for the regeneration of Russia.

Slavophile doctrine – the doctrine of Khomyakov, the two Aksakov brothers, and of the two Kireyevsky brothers – was a heady romantic mixture. This, too, stood in opposition to Nicholas's régime, which it stigmatized as the product of an alien dynasty, stemming ultimately from the European-inspired reforms of Peter the Great. Slavophilism saw in unperverted Russian history a youthful force with its own innate strength and virtue, rooted in the people and the Orthodox Church, destined to supersede the West and to become the universal civilization of the future. In this context, Slavophile doctrine gave special emphasis to the cooperative peasant commune. This embodied, it was argued, a specifically Russian form of socialism of a type that would enable Russia to avoid both the individualism and the unbridled capitalism that were tearing the Western world apart.

The opposition between Westerners and Slavophiles must not be over-estimated. Both shared, albeit for different reasons, a detestation of the régime, and both fervently believed in the Russian future, whether as part of the West or as an independent force. It was Herzen who, in a famous image, compared his allies and his Slavophile opponents to Janus, or the two heads of the Imperial eagle in whose breast the same heart was beating, even though the heads might face opposite ways. Also, as the thirties turned into the forties, events in the West itself helped to bring the two groups together. After 1848 particularly, with the failure of the revolutions in such countries as France and Germany, and the consequent discredit of republicanism and liberalism, the notion of a Russian messianic mission also took root among the erstwhile Westerners. Herzen, who went into voluntary exile in 1847, emerged deeply disenchanted from his first encounter with Western ways, turned a complete *volte-face* and was converted to the Slavophile cult of the peasant commune: '... what a blessing it is for Russia that the rural commune has never been broken up, that private ownership has never replaced the property of the commune: how fortunate it is for the Russian people that they have remained outside all political movements, and, for that matter, outside European civilization ...'[1] Bakunin too began to talk of the Russian peasant as the future progenitor of the revolution. Europeanism lost ground and pro-Russian sentiment grew. Ivan Kireyevsky, in his writings of the early fifties, could look back over the past twenty years and point with satisfaction to the decay of Russian faith in the West.

Among the first fruits of Westernism in a political guise, as distinct from a theoretical attitude, can be seen the circle organized by Petrashevsky in the 1840s in St Petersburg. Butashevich – Petrashevsky himself, to give him his full name, was a graduate of the exclusive *lycée* of Tsarskoe Selo. From there he passed as a junior clerk into the Ministry of Foreign Affairs. He was in his twenties in 1845 when he began to bring together the intellectual élite at Friday-evening gatherings in his apartment at St Petersburg. By now the intelligentsia had lost much of its social homogeneity and Petrashevsky's circle was com-

1. Alexander Herzen, *The Russian People and Socialism – An Open Letter to Jules Michelet*, English translation, London, 1956, p. 189.

posed of 'men of different ranks' – middle-rank officials and army officers, students, artists, and young *littérateurs* such as Dostoevsky, Danilevsky (the precursor of Spengler), and Pleschcheyev, the poet. Here the ideas of pre-Marxist socialists such as Louis Blanc, Fourier, Cabet, and Considérant were read and passionately discussed.

The group had neither formal membership nor common doctrine. But its members were unanimous in their opposition to the autocracy, serfdom, and harsh army discipline. They demanded equality before the law, freedom of the Press, abolition of the censorship, and a democratic republic. These demands were conceived, of course, on the Western models.

But how were they to be attained? The Petrashevskyites condemned the Decembrist solution of a military conspiracy or palace revolution. On the contrary, the revolution could only be successful if it won the support of the masses. 'The main aim', said Petrashevsky, 'is that ideas and wishes should take root in the masses of the people, for when the whole people wants something, then the army will be unable to do anything.' In the Russia of the 1840s this meant a peasant revolution. It was for this reason that Petrashevsky devotedly studied the Peasant Wars in sixteenth-century Germany, the career of Thomas Münzer, and the Levellers' movement in England – not to speak, of course, of the Pugachov rebellion as the Russian version of the Western upheavals.

Under the impact of the revolutions of 1848, these interests and studies took on a more radical character. A plan for an uprising was evolved, or more truly perhaps, the idea of an armed uprising took root in the minds of a few hotheads. One such was Nicholas Speshnev, a wealthy young landowner, a theist, who may have been the original of Stavrogin in *The Devils*. Speshnev said on one occasion

... that the Urals were almost ready to rise, that it was possible to get 400,000 armed men, that this was a terrible force and that the person who controlled it could be almost certain of victory, that ... if anyone with such a force hurled himself into the lower Volga area, where the people remembered Pugachov, then he would be raised on high.

All this soon proved to be empty talk. From early in 1848 the police knew of the existence of the Friday-evening meetings, and by the beginning of 1849 they had planted an informer in the group. In April and May, more than a hundred participants were arrested and interrogated, often after periods of solitary confinement in the Peter and Paul fortress in St Petersburg. In the end, some of the group were sentenced to hard labour and fifteen to death. Among the latter was Dostoevsky. Nicholas played a sadistic joke on these fifteen young men. He did not make known his clemency and the commutation of the death sentence to hard labour until all the preparations for shooting had been gone through – until the prisoners had been clothed in shrouds and the first three victims brought forward.

LITERATURE AS A SOCIAL WEAPON

Uvarov, Nicholas's Minister of Education, is said to have remarked that only when literature ceased to be written would he be able to sleep peacefully. True or not, the remark indicates the fear in which literature was held. That this should be so was largely due to the influence of Vissarion Grigorievitch Belinsky. It was under Belinsky's influence that Russian literature developed its characteristic identification with the life of society. Belinsky created a tradition. No writer or social critic of the Russian nineteenth century withstood his influence.

He was born in 1810 (or 1811) in Finland, the son of a retired naval doctor who later settled down to practise at Chembar (now renamed Belinsky) in the province of Penza. The father was something of a drunkard, and the mother a harassed, irritable, and embittered woman. Neither parent took at all seriously their precocious son's fanatical devotion to literature and to the search for truth. He won a Government scholarship to the University of Moscow, where he almost starved, but studied unremittingly. In 1831 he was expelled from the University for having written a play denouncing Russian conditions, and also, perhaps, because he lacked the necessary solid intellectual background considered essential by the authorities.

Be that as it may, from 1831 until his death from tuberculosis in 1848, Belinsky had to earn his living by his pen. Much of what he wrote was inevitably nothing more than pot-boiling journalism. Nadezhdin, the professor of literature at the University and also editor of *Telescope*, the periodical which had published Tchaadayev's 'Letter', took Belinsky up and commissioned him to write reviews. Philosophically speaking, Belinsky moved from adherence to Hegel, Schelling, and Fichte to sympathy with French and German Utopian anarchists and socialists – Feuerbach, Proudhon, Fourier, Louis Blanc, and Saint-Simon. Crudely and roughly, this was equivalent to the transition from an uncritical acceptance of the *status quo* to a paramount concern for the individual.

It is no exaggeration to say that Belinsky created a new and vastly influential method of literary criticism. A great deal of his achievement is implicit in the famous letter of denunciation with which he overwhelmed the senile Gogol:

The Russian people is right. It sees in writers of Russia its leaders, defenders, and saviours from Russian autocracy, Orthodoxy, and nationality. It can forgive a bad book, but not a harmful one.

Here, in essence, is Belinsky's ideal of the committed writer as the man capable of giving voice to the deepest and most cherished humanist ideals and values. There was a time, as Herzen said, when Germany gave Russia nothing but midwives and empresses. But in the early nineteenth century these yielded pride of place to German philosophy, and Belinsky's ideal is probably derived from German romantic philosophy, with its view of the artist as the mouthpiece and expression of his particular epoch. Under Belinsky's inspiration, however, this developed quite different connotations in Russia.

It involved taking a middle course between the theory of art for art's sake, and moral, social, or political didacticism. Literature would be both narrower and broader in scope – narrower in the sense that its concern would be first and foremost with ideas, and broader in that it would be responsible to the whole of society. 'To deny art the right of serving public interests,' Belinsky wrote, 'means debasing it, not raising it, for that would mean depriving it of its most vital force, that is, the idea, and would make it an object of sybaritic pleasure, a plaything of lazy idlers.'

Thus the critic's task was to elucidate and assess the idea embodied in a work. Belinsky would have agreed that 'all art is propaganda'. The question was: what sort of propaganda? Did it defend the people from 'autocracy, orthodoxy, and nationality', or did it defend the official patriotism of the régime? Where did the writer, the poet, or the novelist stand? This was Belinsky's criterion. In other words, he saw and judged literature in moral and not in literary or utilitarian terms. The work of art had no right to lead an autonomous moral existence in a world divorced from the values of human life. It was part and parcel of the world.

If literature was identified with life, then it followed that there was an even closer identification of the artist with his art. The artist must above all else remain faithful to his vision – and not only when he functioned as an artist but also when he functioned as a man, if indeed the two could be separated. It is this that accounts for the virulence with which writers such as Gogol and Tolstoy, for example, renounced their own earlier writings when they found them incompatible with their later beliefs. They could not look on their works as something apart from their life.

Belinsky's aesthetic, if the word be allowed, had a further consequence. It was a matter of supreme importance to be an artist; for it meant a decision to take an active part in the most momentous struggles of the day. Hence the peculiar strength and the tang of the conflicts that often set one writer or critic against another – Belinsky against Gogol, Tchernyschevsky against Herzen, Pisarev against Pushkin, Tolstoy against Turgenev. What gave these conflicts their special force was not simply personal division or literary theories, but the underlying conviction that the stake was the very future of society itself.

Belinsky's demand that art be coterminous with life, the artist committed to portray reality (though not of course without the exercise of his imagination), the inescapability of a theme or a problem located in a certain milieu known to the writer and recognizable as such by his readers – all this did, of course, gain added strength from the censorship. By prohibiting the free public expression and discussion of public issues, it helped to ensure that fiction would become the favoured medium for debates of this type.

In Belinsky's lifetime the impetus that his theories gave to literature was already showing itself. He died in 1848. But by then Pushkin and Lermontov had created, in Eugene Onegin and Petchorin respectively, the first significant representative of the 'superfluous men' as a criticism of the régime and of social conditions. The type enjoyed a long life and was variously embodied in the works of Turgenev and Chekhov, and of course in Goncharov's Oblomov, the apotheosis of superfluity in the form of apathy. Although differing in detail, they were men whose energy and talent could find no outlet in public service. They were cut off from the court and the régime by their contempt for its values. But they were also cut off from the mass of the people by their superior Europeanized education. Hence they lived and moved in a sort of limbo, animated purely by private concerns.

Eugene Onegin, for example, tells the story of a bored, blasé, and *déraciné* St Petersburg dandy who rebuffs the love of Tatiana, a sincere, well-meaning girl. Onegin flees the country estate where he had met Tatiana, after a duel in which he kills his friend Lensky. He seeks oblivion in the Caucasus. Years later, in St Petersburg, he meets Tatiana again, this time as a mature beauty – but married to an elderly dignitary. Onegin is now in Tatiana's place – it is he who is rebuffed by Tatiana, in her determination to remain faithful to her vows.

Lermontov's *A Hero of our Time* gives full vent to similar frustration. The introduction boldly calls it 'a portrait composed of the vices of all our generation, in their full development'. The author's semi-autobiographical mouthpiece is Petchorin, a young officer whose character is gradually revealed – there is virtually no story to the book – through a series of episodes. Petchorin, *the* man of the thirties, is shown to be something of a Machiavellian character, become sinister in his incapacity to use his undoubted talents and condemned to waste his life in pointless soldiering. Unfathomable and unutterable despair is his end. Although Lermontov's portrait is no objective analysis and not without a certain cool irony, it yet remains a powerful study of a man of vigour gone to seed in a society that denies him the necessary scope. Here the 'superfluous man' is blended with the rebel against society, whom Dostoevsky was later to portray in such characters as Raskolnikov.

Pushkin and Lermontov studied society on a small scale, their viewpoint derived from the position of the isolated Westernized intelligentsia. With Gogol, the canvas is immeasurably broadened to include the life of the small landowner and nobility, as Gogol conceived them, in all its triviality and vulgarity. He himself came from a family of poor Ukrainian gentry. After some years spent as junior clerk, private tutor, and professor of history at the University of St Petersburg, he made literature his career. Pushkin and Zhukovsky, a veteran court poet, were his first mentors.

Gogol's early stories of Ukrainian life, told in a nostalgic, humorous vein, and then his stories of metropolitan St Petersburg – fantasy strangely mingled with big-city realism – soon established him as one of the leading writers of the day. One of these stories, *The Overcoat*, tenderly satirizing and pitying a humble clerk, has become famous as the progenitor of a staple theme of much nineteenth-century literature. It is characterized by a simple story, a meek principal character, an accumulation of small detail, and a tone of fellow-feeling. Dostoevsky's *Poor Folk* is a prime example of the genre. It was also Dostoevsky who said: 'We have all come out from underneath Gogol's "Overcoat".'

With *Dead Souls*, Gogol's peculiar vision of a reality given over to spiritual emptiness reaches its climax. (The 'souls' of the title are serfs, this being the term by which they were known in Russia.) Gogol based the novel on an idea given him by Pushkin. The form is picaresque, the loose thread of the story being held together by the negotiations of the hero, Chichikov, in his quest for the 'dead souls'. His idea is simple. After a number of ups and downs in life, Chichikov, the epitome of smooth imperturbability, decides to enrich himself again by trafficking in the dead – the physically, and not the officially dead. He will buy from the landowner those of his serfs who have died since the last census. This deal will benefit both parties. On the one side it will release the landowner from paying tax on a worthless possession; on the other, it will give Chichikov title to a certain number of serfs whom he can then mortgage to a bank, and thus set himself up as a man of substance once again.

This macabre story is the pretext for a widespread panorama of backwoods noblemen and provincial life. As Chichikov goes about his business of purchasing the dead souls – or rather, 'non-existent souls', as he delicately prefers to describe them – he gives Gogol occasion to evoke a satirically distorted but recognizably truthful picture of country life. This is not the wild Caucasus of Petchorin, not the lyrical estate of Onegin, but a cadaverous world, peopled by barely one sympathetic personage. Whom does Chichikov meet on his travels? Grasping, corrupt officials, miserly widows, slothful, brutal landowners, stupid noblemen, gamblers, and a whole array of living monstrosities. At the end it is clear that the 'Dead Souls' of the title are by no means the serfs but the whole world above them.

When the novel was published in 1842, it created something of a furore. Pushkin had exclaimed some years earlier on hearing Gogol read the first draft: 'God, how sad our Russia is!' Belinsky greeted it as a truthful picture of the country, inspired by a profound inner love. The Slavophiles could see in the novel a faith in the Russian future, however much its present might be betrayed. Gogol's comedy, *The Inspector General*, which is a not dissimilar satire on provincial life, could also be understood in this sense. No doubt this was the reason why Nicholas personally intervened so that it might be presented.

In his later years, Gogol fell victim to a form of religious mania, made a pilgrimage to Palestine, and utterly disappointed the hopes of the liberals and even of the Slavophiles. His last work, *Extracts from Correspondence with Friends*, in which he repented of his earlier writings and praised the virtues of the autocracy and of serfdom, brought upon him the wrath of Belinsky. 'Preacher of the knout,' Belinsky reviled the hapless

Gogol, 'apostle of darkness, champion of obscurantism and dark ignorance. ...' This magnificent torrent of denunciation set the tone for almost the whole of the literature of the nineteenth century. A bad book – yes; a harmful book – no, a thousand times no!

FOREIGN AFFAIRS

It was neither the intelligentsia nor the social tension inside the country that put an end to Nicholas's oppressive régime. It was his own foreign policy. In central Asia and Transcaucasia, Nicholas could advance against the Persians and Turks with relative immunity. In western and central Europe, he could carry out the part of 'the gendarme of Europe' in maintaining the Vienna settlement and meet no opposition. But when he essayed a forward policy in south-east Europe, in the tradition of Peter the Great and Catherine, he came to grief.

All his plans for a carve-up of Turkey, for a weak Turkey under Russian patronage, for the opening of the Black Sea and the Dardanelles to Russian warships, for the extension of Russian influence to the Balkans – all these far-flung aims were shattered by the catastrophe of the Crimean War.

Early in 1853, Nicholas tried to assert more vigorously than ever before the Russian right, as the protector of the Sultan's Orthodox subjects, to intervene in Turkish affairs. But this the Sultan refused to accept, backed up as he was by the French and British. Russia then occupied Moldavia and Walachia; a few months later Turkey declared war. The following year, after the Turkish fleet had been destroyed by the Russians and after vain attempts had been made to compose the differences of the two powers, Britain and France allied themselves with Turkey. Austria too joined the coalition of enemies when it called on Russia to evacuate the two Danubian principalities.

The Allies took the Crimean fortress of Sevastopol in 1855, and after that the Russians had no choice but to surrender. The peace terms put a decisive stop to further Russian advance in the south: certain Bessarabian territories were restored to Moldavia, and this threw the Russians back from the Danube; Russia lost the right to maintain a fleet in the Black Sea; the

Straits were closed to the warships of all nations; and the great powers themselves undertook to safeguard the integrity of Turkey and not to countenance any intervention in its internal affairs.

This was more than a military defeat. It discredited the whole régime. Ill-led and ill-equipped as the British troops were, the serf armies of the Russians had to do and die in incomparably worse conditions, even though they were fighting on their home soil. They had no modern weapons. The serf labour manning the wool, linen, and leather industries proved incapable of supplying any adequate uniforms. A special munitions industry did not exist. The commissariat suffered from the almost entire absence of roads and railways in the immediate hinterland of the fighting area.

In these circumstances, what began as a military trial of strength rapidly developed into a trial of the régime's strength. Many of the intelligentsia looked forward to Russian defeat as an essential pre-condition to reform. In this they were not disappointed.

CHAPTER 11

Reform and Assassination

'IMPERNICKEL is dead! Impernickel is dead!' This cry resounded at Twickenham in March 1855. Small boys were echoing in their own jargon the rejoicing of Herzen, now in exile in London, at the death of the Emperor Nicholas. When he first saw the news in *The Times* he had rushed out in great jubilation to distribute largess to all and sundry.

'Impernickel's' death in itself well justified Herzen's excitement. But there was more. If ever a country existed to which Trotsky's dictum applied – that war is the locomotive of history (especially if the war is unsuccessful) – that country was Russia. Defeat in the Crimea opened the way to impressive political and economic changes. Defeat in 1905 by the Japanese helped to produce the St Petersburg Soviet and the Duma movement. Defeat in 1917 helped to create the greatest upheaval of all, the Bolshevik revolution.

Defeat in 1855 exposed the inadequacy of Russia as rarely before. A social system came to grief as well as a nation. The challenge was taken up. Out of the débâcle came a first great industrial upsurge, a transformation of the countryside; and also the creation of a disciplined party, the Populists, able to gain the upper hand over the police. Professional revolutionaries, unlike such well-intentioned amateurs as the Decembrists and the Petrashevskyites, rose up to challenge the State. In the 1860s and 1870s also, a new realistic, iconoclastic, and utilitarian spirit took hold of the intelligentsia. Much of this flowed from the primary act of the new reign, the emancipation of the serfs. Yet the reign which began with high hopes ended in – assassination. When Alexander II followed his father Nicholas on the throne, he had the wellnigh unanimous support of all groups of Russian society and even of political exiles such as Herzen. A few years before his death, 'the Tsar was isolated from the Russian people, unpopular with the educated public, and cut off from the bulk of society and the Court. His fate had become a matter of indifference to the majority of his subjects.' [1] His death, when it finally came, was something of an anticlimax.

No sooner had Alexander ascended the throne than it became evident that a break with the past was in preparation. In publishing the terms of the treaty of Paris, he also proclaimed that a programme of social reform was imminent. Furthermore, he made it clear that the methods of his father were in abeyance. He released the surviving Decembrists and Petrashevskyites from exile. Many thousands of people were removed from police supervision. He lifted the restrictions imposed by his father on university students, and he put in hand a revision of the censorship regulations. To suspend recruiting, to remit tax arrears, to show more tolerance *vis-à-vis* Poland and the Catholic Church also augured well for the new reign.

The most startling sign of the new order came in an announcement of April 1856, a month after the end of the Crimean War. Alexander was addressing a gathering of the Moscow nobles. He denied rumours that he was intending to put into effect the immediate emancipation of the serfs.

But [he went on] you yourselves are certainly aware that the existing order of serfdom cannot remain unchanged. It is better to abolish serfdom from above than to wait for the time when it will begin to abolish itself from below. I request you, gentlemen, to reflect on how this may be achieved. Convey my words to the nobility for their consideration.

Unwelcome words indeed! But the hostility of the nobles was not uniform, and this gave the Tsar something of a lever to overcome their resistance. There were the so-called planters, as Herzen called them, to whom the land itself, as in the fertile black-soil areas, was of paramount importance; and there were the owners of the less fertile territories in the north and centre supported by the annual *obrok* payments, who were less reluc-

1. W. E. Mosse, *Alexander II and the Modernization of Russia*, London, 1958, pp. 162–3.

tant to part with their land in return for high indemnification payments. In the end the planters' views largely prevailed.

What made sense of the Tsar's initiative was not only the rising tide of peasant unrest,[1] but also the fact that the institution of serfdom was decaying from within. Serf labour in the more advanced branches of industry was already a thing of the past. In agriculture it had long been discovered to be a handicap if large-scale production for the market was to be undertaken. Free hired labour had shown its superiority every time. Finally, and most important of all, in the more industrialized parts of the country, where there were opportunities for employment outside the estate, the serf had already become a sort of disguised workman. He would discharge his obligations to his landlord by paying him *obrok*, a fixed sum in cash, that he had earned in some enterprise. Again, in the black-soil regions, where there existed agricultural over-population, land without serfs fetched a higher price than land with serfs. The landowner had no wish to maintain a superfluous labour force.

Amid this plethora of conflicting interests, Alexander held fast to three principles: the serfs must be released with land; the whole operation must be achieved peaceably; and the former serfs must enjoy full personal freedom from the very day their emancipation was proclaimed. But this still left plenty of room for manoeuvre on the part of the planters, especially since the Tsar turned over the discussion of the proposed reform to provincial committees of the nobility. At the centre stood the main committee, including various editorial commissions and committees of experts, and these examined all the proposals emanating from the provincial committees. It was a unique method of lawmaking for an ostensibly autocratic régime. The Tsar himself, in 1858, undertook a sort of canvassing campaign among the nobility of such towns as Vladimir, Tver, Moscow, Kostroma, urging them to take their full share of the preparatory discussions. It was part of his campaign to split the opposition. Gone were the days when an autocratic fiat carried all before it. At the beginning of 1859 a

1. See pp. 149–50.

start was at last made on drafting the emancipation statutes.
Two years later, after multifarious amendments, they were
signed and promulgated. The whole process had taken close
on five years.

What emerged from this prolonged period of gestation?
Twenty-two enactments in a fat volume of over 300 pages. The
complexity was enormous, but one fact at once stood out: the
emancipation was something of a fraud. This was so deeply
felt that Alexander soon had to deny forcefully rumours of a
new and genuine emancipation to replace the bitter appearance
of 1861. At Bezdna in the province of Kazan it even came to a
massacre by troops of seventy rioting villagers. None the less,
despite – and also because of – its limitations, the statute of
emancipation marked an epoch in the history of modern Russia.
It resounded decisively and perhaps fatally throughout the remaining decades of tsarism. The theme of the reform was the
desire to reconcile every conflicting interest. 'The peasant
should immediately feel that his life has been improved; the
landowner should at once be satisfied that his interests are protected; and stable political order should not be disturbed for
one moment in any locality.' Such was the Government's laud-
able ideal. But it was, in fact, distorted into a concern for the
preservation of the landowners' prerogatives.

The serf, it is true, was at once released from his status and
became a free citizen. He was free to marry, own property in
his own name, take action at law, and engage in a trade or
business on his own account. He could no longer be bought or
sold; but for this he paid a heavy price in economic and financial dependence. This resulted from the principle that the land
belonged to the landowner regardless of whether it had been
used by him or by his serfs. If, therefore, the serf was to be
released with land, he must pay a price for it. Furthermore, the
amount of land and also the type of land that the former serf
received would depend to a large extent on what the landowner
would grant.

To take the latter first – the peasant's pre-reform allotments
were to be regarded as adequate, and voluntary agreement was
to be sought on this basis. But since there were great disparities

among the various regions, as well as within them, maximum and minimum norms were determined for the whole of the country, to which the new allotments had to conform. In most agricultural provinces, however, the owner could retain as much as one-third of his non-waste land, no matter what effect this had on the size of the allotments. There were also 'beggars'' allotments, one-quarter of the maximum norm, which were acceptable only at the peasant's option and for which he had to pay nothing. About three-quarters of a million families accepted these holdings in the hope of bettering their status in the future.

Except for the 'beggars'' allotments, the land allotted to the peasant did not become his personal, inalienable property until it was redeemed. Thus the redemption operation constituted the second main feature of the reform. The land could be redeemed by service at the rate of thirty or forty days' labour per year on the landowner's estate; alternatively – and this was the more general procedure – the *obrok* due to the landowner was capitalized at six per cent. This sum was then known as the redemption value of the allotment. At this point the State intervened between landowner and peasant, paying the former eighty per cent of the redemption value in interest-bearing certificates negotiable on the Stock Exchange, and yielding a six-per-cent income. The remaining twenty per cent was paid, by voluntary agreement, by the peasants themselves. If the settlement was made at the landowner's request, he merely received the certificates to the value of eighty per cent. Whichever was the case, the sums advanced by the Government to the landowners had to be repaid by the peasants over a period of forty-nine years from the date of issue and at the rate of 6·5 per cent to cover interest and amortization.

The village commune was encouraged and strengthened by the reform in that it became collectively responsible not only, as before, for the payment of taxes, but also for the payment of redemption debts. Similarly, in order to preserve the stability of relations in the countryside now that the peasant's personal dependence was ended, a form of peasant self-government was instituted, composed of peasant householders and elected

'Elders'. Special courts to judge minor cases that involved peasants only also came into existence to serve 'cantons', which were formed by the grouping together of communes.

So much for the landowners' serfs, numbering some twenty-one million. The somewhat larger number of State serfs received more generous treatment. The average holding of the privately owned male serf was nine acres, that of the publicly owned serf twenty-three. (This comparison is not, of course, as straightforward as it appears, given the enormous difference in the agricultural value of the land in the different areas – ranging from Siberia to Poland, and from the north to the Caucasus.) Moreover, the publicly owned serf's annual dues were materially less. In other respects, including the attachment of the peasant to the household, commune, and 'canton', there was no essential difference between the two processes of emancipation.

What was the effect of all these measures on the peasantry? The massacre of Bezdna was no isolated event. The Ministry of the Interior reported no fewer than 647 incidents of peasant rioting in the first four months following the promulgation of the emancipation statute; and during the year of 1861 there were 499 serious outbreaks of rioting in which troops had to be used. Looking farther ahead, the terms of emancipation were directly related to the peasant *jacqueries* of 1902, 1905–6 – and the seizure of the land in 1917.

The fact is that in 1861 the peasants actually received less land than they had utilized in the pre-emancipation era; and for that lesser amount they paid more than the land would fetch by sale or rent. In general and summary terms, the peasants lost one-fifth of the land previously held. In many cases this was the most valuable part of the land, such as the meadows and pastures in the north and central areas. In the black-earth zone particularly, where many peasants took the 'beggars' ' holdings, they lost more than a quarter. Only in the non-black-soil industrial provinces did they increase their holdings. As regards payment, the redemption value of the allotment, ostensibly derived purely from the value of the land, did in fact include a hidden element of ransom. The former serf was paying not merely for his allotment; he was also buying from the landowner his personal freedom.

In both respects he received less than he was paying for. It has been calculated that by 1878 only some fifty per cent of the former serfs had received adequate endowments of land. The rest were below subsistence level. This proportion changed even more to the peasant's disadvantage with the growth of rural population, which expanded more than fifty per cent between 1860 and 1897 – from fifty to seventy-nine millions.

In financial terms the landowners did not do much better. By 1860 about sixty-two per cent of their serfs were mortgaged to the banks and about fifty-three per cent of the value of their estates. Thus a large part of the vast sums which would otherwise have accrued to them through the redemption operation actually went to pay off their debts. By 1871, for example, 248 million roubles, out of a total of 543 million paid in redemption dues by the peasants, had had to be repaid to various credit institutions.

The emancipation was not only epoch-making in itself in its effects on rural Russia; it also opened the way to further reforms, which became indispensable now that the most solid bulwark of the old régime had gone down in semi-ruin.

The first reform to follow the emancipation was the institution of a system of local self-government, through an agency known as a *zemstvo* (plural *zemstva*). The members of the *zemstva* were elected on a county and provincial basis through a system of electoral colleges that gave preponderance to the nobility. Those elected then chose an executive board of three members to carry on the routine business of each *zemstvo* from one annual meeting to the next.

Their functions were limited to the maintenance of roads, prisons, hospitals, the encouragement of industry, agriculture, education, poor relief and famine relief, and public health. The *zemstva* were not fully autonomous because they had no executive power, and a good part of their revenue, derived from rates, was used for the purposes of the Central Government. On the other hand, particularly after 1878, as the *zemstva* spread to more and more provinces, they developed into a genuinely liberal force, constantly at odds with the tsarist bureaucracy. What is more, through their employment of experts, that is to say, doctors, nurses, teachers, sanitary inspectors, engineers, economists, and agronomists, they gave employment to a new class of reformers and radicals, intimately aware of peasant conditions and bitterly critical of what they saw.

As a reforming tsar, Alexander II also had to his credit a development of municipal self-government that led to the creation of a primary educational system in the larger towns, as well as an equalization and lightening of the burden of military service by the institution of conscription and the abolition of the more barbarous forms of punishment. Ultimately the period of service was reduced from twenty-five years to fifteen – six with the colours and nine with the reserve.

In the intellectual and educational field there was a relaxation of the censorship of books and periodicals, and an attempt to re-establish university autonomy and widen the basis of entry to secondary schools. These had previously been largely the prerogative of the nobility.

Finally, the early reforming years of Alexander's reign also produced a new court system that provided for open trial, the use of a jury, the appointment of trained judges who were irremoveable, and the right to counsel. Minor offences were reserved for trial by Justices of the Peace elected by the *zemstva* and municipal representative institutions. Again, *toute réserve faite*, especially since the peasants had access only to their own courts and therefore equality before the law could not be said to exist, it still remains true that the legal reforms of the 1860s marked a break with the past as indeed did all the other reforms of the period.

Yet how far did they really go? Did they amount to the 'revolution from above' that Alexander had seemed to proclaim at the beginning of his reign? Obviously not. But it is equally clear that the old régime had gone for ever. Rather, to the existing tensions in the Russian social system was added yet another: that between the autocracy and some intimations of Western-European legal, constitutional, and educational practice. Two worlds came to exist cheek by jowl. The emancipation and the associated reforms powerfully accelerated a tendency that had been gaining strength since the end of the eighteenth century. On the one hand, there was the world of the autocracy, its bureaucratic system, its ecclesiastical appendage, its serf economy, and its quasi-monopoly of intellectual life. All this formed something of an interdependent unity that exhausted virtually every aspect of the national life. There was very little that took place outside the purview of this comprehensive system. Certainly nothing went unsupervised by the relevant authorities, whether it was an independent publisher, a dissident religious sect, or a meeting to discuss the work of a French socialist.

On the other hand, and increasingly so as the nineteenth century progressed, all sorts of new stirrings and new forces were alive. Independence or attempted independence of the State – this was their common characteristic. The most note-worthy was the revolutionary, literary, and intellectual move-ment. The administration of justice was separated from the administration of the State. The Church no longer monopolized education, even at the primary level. The Press took on a character of its own, however much hampered by the censorship. And the *zemstvo* movement, despite the supervision exercised by the Government and its inherent weakness in financial and executive power, constituted a source of authority directly rep-resentative of the population.

All this signified not only the clipping of autocratic wings, but also the rise of new centres of power and influence, opposed in varying degrees to what remained of the old régime.

The friction resulting from the juxtaposition of these two worlds was intensified by the economic consequences of the reforms and the increased industrialization of the 1860s and 1870s.

ECONOMIC DEVELOPMENT

Both the landowners and the peasants lost by the reform. Many of the former saw their mortgage indebtedness reach a figure exceeding the pre-reform total. Many a Chekhovian cherry

orchard had to be abandoned to the merchant, emergent indus-
trialist, or wealthier peasant. The nobles' landholdings dwindled
inexorably; they shrank by something like eighteen per cent in
forty-one provinces of European Russia in the two decades
following emancipation, and at an increasing rate afterwards.

On the whole, the peasantry fared even worse in that their
standard of living actually declined. The most unfortunate of
all, of course, were the four million peasants of both sexes who
received no land at all. These were predominantly the former
domestic serfs, the serfs employed in manorial factories and
mines and the State industrial installations. For the majority of
these there was no choice but to drift townwards.

Their counterparts who remained on the land fared little
better. The terms of emancipation in themselves ensured that
the average peasant-holding was less than it was before 1861.
Even this inadequate amount diminished with the growth of the
rural population. Land could be rented for additional cultiva-
tion, it is true, but this required cash resources beyond the
peasant's purse. In the 1880s, something like one-quarter of the
peasants did not even own a horse. Furthermore, emancipation
dues and other taxes produced a steadily increasing peasant
indebtedness. It was this, in fact, that enabled Russia to maintain a consistently favourable balance of trade, to service all its
foreign debt in the second half of the nineteenth century, and
to maintain a strong rouble backed by gold. In order to pay his
debts, the peasant had to sell his grain for export to Western
Europe. But 'the granary of Europe' achieved its position only
at the price of impoverishing its own population. The economic
pressure was such that the peasant had to sell the grain that he
actually needed for his own consumption. Chronic under-
nourishment in the Russian village was emphasized by the dis-
proportionate effects of crop failure and, even more, by the
increased mortality rate – from twenty-four to twenty-seven
per thousand at the beginning of the nineteenth century to
thirty-five per thousand in 1880.

By and large, the Russian masses completely failed to participate in the rise of the popular standard of living that elsewhere
characterized the latter part of the nineteenth century. On the

contrary, the acquisition of some degree of personal freedom was accompanied by a decline in living-standards, by land hunger, and by rural over-population.

This extremely low standard of living severely inhibited the growth of the Russian internal market. None the less, emancipation did give a certain impetus to capitalist development and to the urbanization of Russia. After the reform, for example, the urban population increased almost twice as fast as the rural. Again, during the 1880s the first sizeable influx of foreign capital took place.

The development of the railway network was one of the most impressive tokens of the post-reform decades. The aim was first to link the grain-producing areas with the population centres and the internal markets, and second, to link the former with the ports and thus facilitate the export of grain. Between 1861 and 1880 the track grew from 1,000 miles to more than 14,000 miles. Here was something akin to the English railway boom of the 1840s. Both foreign and native interests put up the capital. Russian industry was never able to supply all the rails, locomotives, and other capital items that this development needed. Even screws and bolts had to be imported from abroad. Yet the railway boom did encourage Russian heavy industry, for it constituted the latter's chief market. This can be seen in a sixteenfold rise in coal production, a tenfold rise in steel, and a fifty-per-cent rise in iron and pig-iron in the period from 1860 to 1876. To this period, for example, belong Hughes's factory for the production of coal, iron, and rails to the Krivoi Rog basin, the first large-scale exploitation of the coal deposits in the Donetz basin, and the oil companies financed by the Nobel brothers in Baku. In the traditional industries, cotton-spinning, and woollen manufactures, similar expansion took place.

The fairs served as the great medium of circulation for the products of light industry and consumer goods. At these colourful assemblies, of which there were more than 6,500 in the middle 1860s, agricultural raw materials were collected and industrial products distributed. The fairs made up for the virtual absence of any retail network. The development of railways, the spread of money relations in the countryside, and the undermining of household crafts, turning the peasant into a purchaser of manufactured goods – all these factors promoted the prosperity of the fairs. Their turnover grew from 360 million roubles in 1860 to 460 in 1863.

Concomitant with all this development went the inauguration of a budgeting system (before 1862 there had been no proper budget at all); the promotion of private banks; and various attempts to stabilize the currency. None of these really succeeded, and their failure is evident in the fact that to service the public debt in 1860–80 took up between one-quarter and one-third of all budget expenditure.

As Russian capitalism developed, it began to display certain individual features that set it apart from the Western form. Of course, there were also many resemblances. The exploitation of the labour of women and children, low wages, an extensive system of fines, payments in kind, workers' barracks – this was the same harsh régime as in the early stages of the Industrial Revolution in England. But what first set Russian industrial development apart was the high degree of large-scale production. In respect of organization, Russian industry was as advanced as that of any Western state, if not more so. Thus, forty per cent of the workers, according to calculations made by Lenin, were employed in factories employing more than 1,000 workers. This was in 1879; in 1866 the comparable figures had been twenty-seven per cent. Over the same period the proportionate number of workers in factories employing between a hundred and 999 actually dropped.

This concentration of production was further expressed in Russian industry's characteristically close connexion with the State. This had been a feature ever since the days of Ivan the Terrible, and, even more, of Peter the Great. It was the State that had made serf-workers available, had supplied a not inconsiderable portion of the finance, had leased out the concessions, and had also provided the markets. This is connected historically with the failure to develop an industrial and capitalist bourgeoisie, as distinct from a professional and intellectual one. The typical Russian bourgeois was not an independent entrepreneur or manufacturer, but some kind of professional man, bureaucrat, or administrator.

Given this situation, the State stepped in as the promoter of industrial enterprise. Its instrument was the State Bank, founded in 1860. The issuing of notes was a very secondary function; far more important was its role as the keystone of the Russian credit system. The banks, with the State Bank at their head, were 'not only the creditors of many industrial enterprises but also their founders, cashiers, directors of their current accounts, the managers and owners of their shares and their basic capital, in short, the complete masters of their destinies'.[1]

Naturally enough, therefore, it was the needs of the State rather than the profits of private capital that were the determining factor, though not of course the *exclusive* factor, in laying down the general lines of industrial development. This did not mean that the class struggle was waged any the less fiercely. This is best exemplified in the history of Russian railway construction. From the start, the State's participation was extensive and its impetus decisive. The first railway company resulted from an Imperial decree of 1857 establishing a Grand Company of Russian Railways, formed of an alliance between the State and a consortium of French, Dutch, and British bankers. The State granted the concession and also determined the direction of the network. This would be operated by the Company, which would also receive a State-guaranteed five-per-cent interest on the sums invested. Later, owing to the Company's difficulties, the State took on a more direct role in its fortunes. The headquarters were transferred from Paris to St Petersburg and four Russians were appointed to the board of management.

This set the general pattern of participation between the State and private, mainly foreign, capital. In some cases the State would even advance the capital sums required for the construction of the lines. The logical consequence of this policy was reached in the years after 1876 when the Government began to buy back the railway shares issued to the public and encouraged the consolidation of the smaller companies into large groupings. This was the first step towards the formal

1. Lyashchenko, op. cit., p. 708.

nationalization of the railways. By 1901 the Government owned two-thirds of the existing track. This type of governmental economic policy has an obvious affinity with the planning from above practised by the Bolsheviks.

FOREIGN AFFAIRS

Alexander II and his Foreign Minister Gorchakov had a compelling need to pursue a policy of peace and quiescence. The Crimean War had shown that the country was, in any case, unfitted to wage an extended campaign. Moreover, the reforms of the early years of the reign entailed retrenchment abroad. Lastly, the Polish revolt of 1863, which took more than a year to suppress, was a reminder of the poisons liable to erupt at any moment in the Russian body politic. What with the Italian liberation movement, the struggle for hegemony in Germany between Austria and Prussia, and the ever-present Polish nationalism, it is clear that Russia's best policy was to be pacific, for the maintenance of the *status quo* would offer the best hope of insulating the Empire from the spirit pervading the Western world. The impact of liberalism, nationalism, and self-determination could not but jeopardize the integrity of the Empire. For the time being, therefore, it was politic to suspend the rectification of the losses suffered as a result of the Crimean fiasco, such as the neutralization of the Black Sea and the lost portions of Bessarabia. It was the best part of twenty years before Russia again seriously pursued a forward policy *vis-à-vis* Turkey and the Balkans.

In central Asia and the Far East it was a very different story. These were the happy hunting-grounds of Russian imperialist adventurers, dubious carpet-baggers, and pseudo-viceroys. Trade followed the flag. If European Russia was becoming in some sense a colony of Western capital, then central Asia was also in some sense a colony of European Russia. By Alexander II's time great inroads had already been made into the Caucasus; and by 1864, after further campaigns against the mountaineers, the two-headed eagle was dominant in the territory between the Black Sea and the Caspian.

It was not long before the Caspian was crossed and a base secured for further penetration south-eastwards. This was one axis of advance. In 1869 it led to the foundation of Krasnovodsk on the eastern shores of the Caspian, and from there to the capture of Ashkabad and the Merv oasis in the region of the Afghan frontier. This process was barely completed when military engineers were at work building the Trans-Caspian Railway along the Persian border. This line was later extended southwards to Kushk, on the Afghan frontier. All this area was developed as a cotton-growing centre for the Moscow textile industry.

Meanwhile, farther to the north and east, other Russian forces were coming to grips with the ancient Muslim khanates and emirates of Khiva, Bokhara, and Kokand. Orenburg, on the Ural River, was the centre of Russian expansionism here. Perovsky, the Governor-General of Orenburg, made his first assault on Khiva in 1839. But inclement weather foiled his Cossack cavalrymen. For the next two decades or so, Perovsky conducted a campaign of subjection by building lines of forts deeper and deeper into the tribesmen's territory. By the 1860s they extended from the mouth of the Syr-Darya River (Jaxartes) to Verny, not far from the border of Chinese Turkestan (Sinkiang). Frequent clashes between the tribesmen and Russian invaders along these ill-defined borders gave the occasion for further and more successful Russian aggression. Tashkent fell in 1865; Samarkand in 1868. By 1882, Bokhara, Khiva, and Kokand had all been annexed by Russia and incorporated in the newly formed Government-General of Turkestan. The Russian frontiers were now contiguous with those of Persia, China, and Afghanistan. A railway running south-eastward from Orenburg to Turkestan, where it linked up with the Trans-Caspian line, consolidated the conquest, and made it possible to deploy troops at speed in these far-flung corners of the Empire.

This Russian expansionism often cut across the pacific policy enjoined in St Petersburg. And this urge on the part of a Perovsky in Turkestan, of a von Kauffmann in Transcaspia, of a Chernyayev in Tashkent, created at times profound Anglo-

Russian tension. Only Afghanistan, where British and Russian envoys jostled for position, separated the two Empires. Would the future see a revival of the Emperor Paul's plan of 1801 to invade India? Such were the thoughts and fears aroused in British circles by the Russian advance.

In the Far East the coast was clearer. At first the status of the island of Sakhalin gave rise to friction with Japan. But in 1875 it was agreed that Japan would renounce its claim to the island in return for the Russian cession of those Kurile Islands in its possession. This exchange put Russia in a strategic position at the mouth of the River Amur.

At first, policy *vis-à-vis* China lay in the forceful hands of Muravyov-Amursky, appointed Governor-General of Eastern Siberia in 1847, and then of Admiral Putyatin and General Ignatiev. None of the three paid much attention to the directions emanating from remote St Petersburg, and not much more to the treaties signed at various times between China and Russia. Muravyov himself extended the Russian area of occupation and penetration at the mouth of the Amur River and along its northern bank. This area was later reorganized as a new Russian province and the Chinese were forced to accept it as a *fait accompli.* China came simultaneously under attack from France and Britain over Canton and Tonking, and this also helped the Russians forward. Putyatin and Ignatiev were able to force the Chinese to grant Russia the same trading rights won by the Western Powers and the right also to maintain a legation at Peking. Territorially, Russia acquired land along both banks of the Amur and a strip of coast off the Sea of Japan. Here, in 1861, was founded the naval base fittingly named Vladivostok – Commander of the East.

The other important event in the Pacific at this time was the Russian sale of Alaska to the United States for $7,200,000.

To return to Europe – and the perennial Eastern question. A somewhat similar situation confronted the Tsar as in central Asia and the Far East. Again the nominal autocrat showed himself unable to master aggressive expansionist forces. In Asia, it was the Russian version of the white man's burden; in Europe, Pan-Slavism, a programme of imperialism masquerading as an ethnographical and historical theory. Moreover, Russia, like the other European powers, found that along with their gradual extra-European expansion, matters of foreign policy could no longer be divided into separate geographical areas. What happened at one point on the Russian periphery affected and was affected by what happened elsewhere. There was no longer one single identifiable Eastern question to harass Anglo-Russian relations, for example, but a whole string of interrelated problems ranging all the way from Tashkent to the Black Sea. It was part of the process whereby the impact of the conquering West made the world one. Russia, therefore, in its efforts to expand, would inevitably have to meet the same hostility at any point on its borders. On the other hand, it would also be aided by the conflicts among its neighbours. The success of Russian encroachment on China owed not a little to this factor.

In Europe, the same factor gave Russia the opportunity to overthrow part of the burden imposed by the treaty of Paris. The Tsar and Gorchakov were agreed, while far preferring a pacific policy, that the favourable conjuncture of circumstances present in 1870 was too good to miss. France had collapsed under the Prussian attack, and Prussia itself was to some degree indebted to Russia for the latter's part in preventing Austrian intervention. In the autumn of 1870 Gorchakov could, with impunity, send to the powers who had signed the Treaty of Paris a note repudiating its Black Sea provisions. At first there resulted a Russo-British diplomatic crisis. But there was no alternative to a war against Russia if Gorchakov's unilateral repudiation was not swallowed. A conference in London in January 1871 rubber-stamped this solution. Henceforth both Turkey and Russia enjoyed the right to maintain warships in the Black Sea and to erect fortifications and arsenals on its shores.

This proved to be the first step in a grandiose Pan-Slav campaign. Two books had special influence – Fadeyev's *Opinion on the Eastern Question* and Danilevsky's *Russia and Europe*. Both argued, in summary terms, that the Russian historic mission was to form a federation of the Slav peoples through their liberation from the dominion of Turkey and Austria-

Hungary. Given the repression exercised by Russia in Poland and the Ukraine, this programme had unwelcome and even revolutionary implications. For this reason Pan-Slavism never won the Tsar's favour. But it had such great appeal to influential circles of Russian society – primarily militarist and the official intelligentsia – that it acquired a momentum of its own, overriding the autocrat's hostility.

On the other hand, the Pan-Slav appeal undoubtedly fitted into the framework of traditional Russian policy in the Balkans, and this gave it a measure of respectability. The test came in 1875 when revolts broke out against Turkish rule in Bosnia and Herzegovina. The area of unrest gradually spread as rumour, atrocity, Pan-Slav conspiracies, and excited public opinion made their own specific influence felt. In the summer of the next year, with General Chernyayev, the victor of Tashkent, at their head, Serbian and Montenegrin forces took the field against the Turks.

This new development brought to Moscow and St Petersburg a hectic and hysterical atmosphere. Volunteers in their thousands came forward to join the Serbs; collections were organized; sermons delivered; Press campaigns launched. But at the beginning of 1877 the Serbs were quelled and forced to seek peace. Now, at last, the Russians themselves openly declared war on Turkey. They launched a two-front attack – in the Balkans, based on Romania, and in Transcaucasia, where the Russo-Turkish frontiers met. Again, it was a war between the 'one-eyed and the blind' – so many errors of strategy and judgement were committed.

Russia gained a military victory over the Turks, but it partially forfeited this through diplomatic isolation. At first, the Russians imposed the favourable Treaty of San Stefano. But when a British squadron anchored just off Constantinople, when Bismarck declared German neutrality, and when Austria-Hungary appeared to stand on the brink of war, there was no choice for Russia but to allow San Stefano to go forward for revision. The result was the Treaty of Berlin of 1878. This confirmed the independence of Serbia, Montenegro, and Romania. It also confirmed the Russian annexation of Bessarabia, and of

Kars and Ardahan in Asia Minor. But the biggest difference between the two treaties lay in the reduction of Bulgaria, the southern part of which was restored to the political and military authority of the Turks. The second major difference was the allocation of Bosnia and Herzegovina to the administration of Austria-Hungary. These arrangements, time would show, brought no stability to the Balkans. The multiplicity of compromises gave fresh impetus to nationalist passions and created fresh opportunities for the great powers to back their respective clients.

LITERATURE AND REVOLUTION

The missionary idea at the heart of Pan-Slavism had its revolutionary counterpart. If Danilevsky could say that Russia would set an example to the world by creating a united Slav civilization, then for their part, the radical intelligentsia saw another instrument of salvation in the peasant commune. This cooperative form of landholding would preserve Russia from the bane of a landless proletariat, inhibit the growth of individualism, and herald the birth of a Russian socialism *sui generis*. The virtues of the commune were a staple theme of the thought of the 1860s and 1870s. The new age was also utilitarian, positivist, realist, and practical. The function of art and literature underwent a certain vulgarization, as compared with Belinsky's theories, and was redefined as the service of society. Values and morals became a creation of the needs of society. The noblest calling was that of the practitioner of some branch of the natural sciences. The urge that service to society constituted the true and ultimate aim of life dominated all minds.

A passage from Dostoevsky's *The Devils* gives some idea of the intellectual ferment of the 1860s:

They talked of the abolition of the censorship, and of phonetic spelling, of the substitution of the Latin characters for the Russian alphabet ... of splitting Russia into nationalities, united in a free federation: of the abolition of the Army and Navy, of the restoration of Poland as far as the Dnieper, of the peasant reforms and of the Manifestoes, of the abolition of the hereditary principle and of the family, of children, of priests, of women's rights.

In the end, what crystallized out of all this was the somewhat vague doctrine of Populism. It has as its *point de départ* the conviction that a revolution must come, that the revolution would be socialist and that its institutional kernel would be the peasant commune. The driving-force behind all this would be the people, conceived of as the peasantry. This was a throw-back to Slavophile teaching.

Populism was not only vague; it also suffered from a basic dichotomy, between self-assertion and self-abnegation. Could the world be best remade through the individual's capacity to lead and inspire; or should not the individual seek to sink and steep himself in the mass, transforming it from within? These were the two poles of agitational activity to which corresponded two types of political activity: terrorism, or pilgrimages *en masse* to the people. To this there further corresponded two views of the State: was it a Jacobin instrument of coercion or a retrograde survival that must be done away with, in a spirit of anarchism?

Who were the intellectual leaders of the new decades? At the end of the fifties and the beginning of the sixties, Herzen and the Russian journal that he published in London, *The Bell*, won an extraordinary circulation in Russia, dominating the thought of the younger generation. But Herzen lost much of his influence when his support for the Tsar was undermined by the terms of the Act of Emancipation and also through his support of the Polish revolt of 1863. But to the Populist movement he left his faith in the commune as a Russian form of Socialism that would preserve man from both the all-powerful State and the perils of individualism, and two slogans of immense historical resonance: 'To the People' and 'Land and Liberty'. Whether Herzen would have approved of the use to which they were put is a very different matter. The future belonged to revolutionaries of a more practical and narrower stamp and outlook. They were less of the *grand seigneur* than Herzen, but more in touch with the realities of political life and more prepared to bend themselves to its exigencies in the form of organization.

In the sixties the iconoclast *par excellence* was the young

Dmitri Pisarev, scion of a gentry family. He scorned all dogma and authority, all aesthetic and metaphysical theory, all religion and morals. The voice of reason and the principles of natural sciences were man's sole guide to conduct. To be a 'thinking realist' represented the highest ideal. This led to the most obvious kind of utilitarianism. Down with the romantic view of the artist as an inspired creator, exulted Pisarev! A poet was nothing more than a craftsman. Better be a shoemaker than a Raphael, for at least the shoemaker created something useful. Nature was not a temple to be revered but a workshop to be dominated. Music, painting, and the arts were illusions, only to be savoured by aesthetes blind and deaf to the human suffering around them.

This doctrine was too individualistic to be fully acceptable over a period. But its inculcation of the ideal of service to the community fitted in with Populist thinking.

Further impetus, in the same sense, came from Tchernyshevsky, the son of a priest. He was born in 1828 in Saratov and educated in an ecclesiastical seminary. He rapidly threw off any dogmatic influences and emerged as one of the leading intellectual mentors of the day, radiating an influence that survived twenty years' Siberian exile. Although primarily an economist, what gave Tchernyshevsky his influence was his vastly popular didactic novel *What is to be Done?* (1863). This was actually written during the author's imprisonment in the Peter and Paul fortress in St Petersburg. The directing ideas are taken from Robert Owen, Fourier, John Stuart Mill – whose *Principles of Political Economy* Tchernyshevsky had translated into Russian – and William Godwin, the English anarchist. The theme and message are purely Russian, however. The heroine, Vera Pavlovna, the new emancipated woman, studies medicine and runs a cooperative of seamstresses. Kirsanov, the hero, is a free-thinking physician. Rakhmetov, an aristocrat, who has gone over to the people and devotes himself unreservedly to their welfare, is the real key to the novel. He sleeps on a bed of nails to develop his powers of endurance; he will eat no food but what would be consumed by the poorest peasant; he will renounce all pleasure for the sake of the cause. However lifeless in art, in real life Rakhmetov became the prototype of many a self-sacrificing young person in the revolutionary movement.

The philosophic sanction for self-sacrifice by the more gifted in the cause of the humble came from Professor Peter Lavrov, a teacher of mathematics at the Artillery School in Moscow. His *Historical Letters* appeared in 1868–9 and won even more popularity than *What is to be Done?* He counselled his readers to abstain from the direct political struggle but to stake their all on the social revolution, but he did not exhort them to violence. The objective conditions of revolt already existed in the misery of the masses. What was needed was to awaken the masses to the subjective awareness of their misery, and this was the duty of the educated minority. They had acquired their education and culture at the expense of the downtrodden, and now they must repay their debt. They must show the masses the way to a new social order that would transform their status. And by so doing, the minority would at the same time be fulfilling its true function. This was the rationale of repentance.

As always in Russia, intellectual ferment both erupted into and was fed by the novel. The themes of the intellectuals *vis-à-vis* the peasantry are met with, for example, in Leskov's topical novels, such as *Nowhere to Go* and *At Daggers Drawn* and also in Gleb Uspensky's *Power of the Soil*. But it is in the work of the major novelists of the period – Turgenev, Dostoevsky, and Tolstoy – that the full force of Populist themes and types is dominant.

Turgenev's work, in particular, can be seen as a spectrum reflecting the whole gamut of intellectual evolution. In his first long novel *Rudin* (1855) he had portrayed the superfluous man of the forties – a *beau parleur*, an idealist, a romantic with no firm footing in reality. In *On the Eve* he showed Insarov, the strong man, who is significantly not a Russian but a Bulgarian, preparing to fight for his country's liberty against the Turks. It is typical of Turgenev's pessimism and detachment that Insarov, who is tubercular, never gets to Bulgaria but dies in Venice. It is Elena, Insarov's Russian fiancée, who takes up the dead man's work.

In *Fathers and Sons* (1862), with the creation of Bazarov,

Turgenev went one step further and depicted a Russian man of action, representative of his generation, the man of the sixties. Bazarov is a doctor, a realist, and a practical man who shuns all transcendentalism and whose aim is the mastery of nature. This was the novel that launched the term 'nihilist' with its celebrated definition: 'A Nihilist is a man who does not bow before any authorities, who does not accept any single principle on trust, however much respect surrounds this principle.' But in the end the self-styled nihilist and would-be master of nature himself falls victim to nature. Bazarov dies of typhus contracted while attending the stricken peasantry.

This ironical conclusion and the author's love-hate relationship to Bazarov brought down on Turgenev's head the wrath of both camps. To the men of the left, Bazarov was a caricature of their ideals; to those of the right, a monster, an upstart *révolté*. The furore testified to the acuteness with which Turgenev had discerned a type. In later works his ambivalence to political trends lost some of its detachment, and in *Smoke* and *Virgin Soil* he went over to the direct satire of all Populist types. In the end this coalesced with a sense of the futility of action and of the unreal, impalpable, evanescent, smoke-like nature of life, especially Russian life: '. . . Everything is hurrying away, everything is speeding off somewhere – and everything vanishes without a trace, without ever achieving anything. . . .'

Neither Tolstoy nor Dostoevsky ever identified his work with any special political theme, in the same way that Turgenev did. But since their major novels appeared in the sixties and seventies, and since neither was at all aloof from contemporary topics, it was inevitable that their works would reflect some aspects at least of Populist theory. In Tolstoy this took the form of an idealization of the virtues of rural life and of its chief exponent, the peasant. In the Rousseauesque spirit, a consistent theme of Tolstoy's writing was the opposition of nature and civilization, the former alone inspiring life and fertility. Who is it, for example, who saves Levin in *Anna Karenina* from spiritual torment and suicide but an illiterate unreflecting peasant? Who is it but the peasant Platon Karataev in *War and Peace* who incarnates a deeper wisdom and greater power than any general or thinker? This is precisely because the peasants have not been subjected to the debilitating effects of civilization and can thus, *en masse*, act as a regenerating historical force.

Dostoevsky took his stand at the other wing of Russian messianism – the Pan-Slav movement with the Russian Army as its instrument, its deification of the autocracy, and its invocation of God and the Orthodox Church. Here was 'a true preacher of the knout', to borrow a phrase from Belinsky.

Dostoevsky, like the revolutionaries, could talk of the Russian people's mission to rescue the world from the fate that had overtaken the West. But when Dostoevsky wrote, 'God will save Russia. Salvation will come from the people, from their faith and their humility,' this was no Populist panegyric. This was a call to throw off the sins of nihilism and to emerge as the only God-fearing and God-bearing nation of Europe, in fact as the Pan-European nation *par excellence*. When two such missions clashed, only one could survive. And the Orthodox Church, the autocracy, and the cause of Holy Russia were Dostoevsky's response to the mission proclaimed by the enemy. If, as Dostoevsky averred, 'God will save Russia,' then there was no room for a competing saviour such as Tchernyshevsky, a philosophical materialist who denied God.

'TO THE PEOPLE'

'The road of history is not the pavement of the Nevsky Prospect. . . . He who fears dirty boots must not take part in public activity.' This dictum of Tchernyshevsky proved itself again and again in the course of Russian history. At times it even seems as though a state of latent civil war was characteristic of Russian public life – waged now against dissident nationalities such as the Poles and Ukrainians, now against dissident religious groups such as the Old Believers, now against selected political groups such as the Populists, and now against selected social groups such as the peasants and students. All this had its counterpart, to a greater or lesser degree of severity, in any European country. But it was in Russia that the tension seemed to switch with unique rapidity from one affected area to another.

In the sixties violence was only sporadic and organized on a relatively minor scale. It was marked by a great deal of student unrest that led at times to the closure of the universities, to the temporary deportation of the ringleaders, and the opening of a 'free university' in St Petersburg by some nonconformist, socialist-minded professors.

At the same time illegal pamphlets, appeals, and proclamations appeared (*The Great Russian* and *Young Russia* were two of the best-known series – and a new legal publication, *People's Chronicle*), to give voice to the revolutionary aspirations of the day. It was symptomatic of the contemporary mood that the latter made no mention of the death of the Grand Duke Nicholas in 1865, whereas that of Abraham Lincoln, two days later, received prominent front-page treatment, surrounded by a mourning border. This somewhat inchoate movement produced the first political trial in Alexander II's reign – a landmark at a period when the reforms were intended to inaugurate a new mood between the autocrat and his subjects. It was the trial of Mikhailov, a poet and part-author of a famous *Proclamation to the Younger Generation*. He was sentenced to penal servitude in Siberia, where he died. Pisarev was also arrested at this time; and Tchernyshevsky was stilled for ever. A sentence of fourteen years' hard labour in Siberia and six years' exile left him a broken man when he eventually returned to European Russia.

The radical aspiration of the sixties came most clearly to the fore in a secret ascetic group of terrorists known as 'Hell'; and in another group organized by Nechayev. Both derived a great deal of their inspiration from the Anarchist, Bakunin. The first produced the sickly and ill-balanced student Karakozov, who made an unsuccessful and unauthorized attempt on the Tsar's life in April 1866. Nechayev's importance comes from the cell type of organization which he pioneered, and also from his collaboration with Bakunin in the compilation of the famous *Catechism of the Revolutionary*. This demanded of every member of the cause the uttermost subjection of the personality.

The notorious Ivanov case showed how literally this must be taken. Ivanov, a student at the Agricultural Academy in Moscow and a member of one of Nechayev's cells, had a somewhat independent mind and even doubted at times the existence of a mysterious Central Committee in whose name Nechayev allegedly acted. This 'treason' to the cause – which was in fact perfectly justifiable since the Committee had no existence – provoked Nechayev into organizing Ivanov's murder. In this he involved the other members of his cell. Nechayev died after ten years' imprisonment.

Dilettantism of this type proved its own undoing, at least for a time. Not only did the revelations of the Nechayev trial discredit the radicals, but their impotence was all too clear. There was a reversion to propagandist activity of a more peaceful type. This was at the beginning of the seventies.

In the meantime, the Government made a number of tactical errors which had the effect of alienating further its opponents. First, after renewed outbreaks of student unrest, it expelled large numbers of young men from the universities, depriving them of their professional aspirations, leaving them without career or subsistence, and virtually giving them no hope save for a transformation of society. Second, women students abroad were ordered to return home in 1873 – but not before they had absorbed the teachings of the exiled Russian revolutionaries, especially in Switzerland. Here were more recruits to the cause.

The reaction, not to say revulsion, from the conspiratorial, underground, and explicitly political methods of Nechayev led to the formation of a number of non-conspiratorial student circles. Informality and loose organization were their characteristics. Nicholas Chaikovsky and Mark Natanson founded two of the first circles in St Petersburg in 1869. Others spread, to Kiev, Moscow, Kazan, and Odessa. Their members came from the intelligentsia of the upper and middle classes, with a sprinkling of army officers. One girl, Sophie Perovskaya, for example, was the daughter of a former Governor-General of St Petersburg. Prince Peter Kropotkin, the future revolutionary anarchist, of ancient Russian lineage, was another member.

The primary purpose of these circles was self-education and

self-betterment through study. The next step was to extend this process to the less privileged – the urban workers and peasants. The idea was to draw in the popular masses.

Gradually they came to the idea [writes Kropotkin] that the only way was to settle among the people and to live the people's life. Young men went into the villages as doctors, doctors' assistants, teachers, village scribes, even as agricultural labourers, blacksmiths, woodcutters, and so on, and tried to live there in close contact with the peasants. Girls passed teachers' examinations, learned midwifery or nursing, and went by the hundred into the villages, devoting themselves entirely to the poorest part of the population.

In Moscow, writes Stepniak, another participant in this movement, girls 'become common mill hands, wrought fifteen hours a day in Moscow cotton factories, endured cold, hunger, and dirt ... in order that they might preach the new gospel as sisters and friends, not as superiors'.

Illegal libraries were set up, books and pamphlets distributed illegally and published illegally – the works of Marx, Proudhon, Lasalle, Lavrov, Louis Blanc. This all took place in the name of Lavrov rather than that of Bakunin. Patient preparation and patient sowing of the socialist seed, conceived in a social not a political sense, must take precedence over direct revolutionary action. Besides, the crushing of the Paris Commune in 1871 served to reinforce the view that armed insurrection was foredoomed to failure and that a specifically Russian way must be found.

The opportunity came in the spring of 1874. The previous winter there had been a famine in the Volga region. This brought the ever-present sense of guilt to a new pitch of acuity. The result was an extraordinary phenomenon. Between two and three thousand of the intelligentsia, perhaps a quarter of them women, literally 'went to the people'. In a spontaneous elemental upsurge they put on peasant dress and tried to become one with the peasantry. To the crusaders it was not so much a political agitation, Stepniak wrote, but 'far more a mystical and religious movement'. The élite of Russian youth threw aside their careers, positions, and privileged past in order to spread the gospel of socialism.

All in vain. The peasants refused to listen to the gospel of the crusaders, preached in what Kropotkin called that 'mad summer' of 1874. No point of contact developed. Neither side understood or talked the same language as the other. Never was the dichotomy between the outlook of the intelligentsia and educated classes and that of the 'dark masses' more dramatically revealed.

But what the peasants failed to understand, the Government did. It treated the movement as an attempted revolution. By the end of the autumn the police had their hands on more than 1,500 young people. Further arrests were made when a second 'going to the people' took place in 1875. This was as much of a fiasco as the first.

Two mass trials were staged – the 'Trial of the Fifty' in Moscow in the early part of 1877 and that of the 'Hundred and Ninety-Three' in St Petersburg in the autumn of the same year. The sentences were comparatively light, and most of the accused were acquitted. But there was no sign of any concession by the Government to the Populists' demands. In these circumstances, there was a reversion to the type of theory and activity more popular in the sixties – that of a disciplined group with a quasi-terrorist outlook.

Through Natanson's inspiration in the main, the famous organization 'Land and Liberty' was founded in 1876. It demanded that the land be handed over to the peasants and that the State be destroyed in the name of collectivism. What this programme lacked in clarity it made up in organization. Here was a close-knit, disciplined, underground body, organized in regional groups with a membership of not much more than 200, with its own printing press, and network of secret sympathizers and fellow-travellers. A special 'disorganization group' existed to protect members from the police and to plan escapes if necessary. 'Disorganization' also included the assassination of informers and prominent members of the opposition. 'Land and Liberty' acted in the name of the people, in the true Populist tradition. But it had given up hopes of actually calling in the aid of the people, although, of course, it might and did exploit and encourage all symptoms of popular discontent. Thus one of its first acts was to stage a mass demonstration in front of the Kazan Cathedral in St Petersburg. More than 2,000 people saw and heard a young student of the St Petersburg Mining Academy, George Plekhanov, deliver a rousing speech and unfurl a red flag bearing the words: 'Land and Liberty'. Plekhanov, a decade or so later, played an invaluable role in introducing Marxism into Russia.

The same demonstration was also important in giving, indirectly, powerful encouragement to terrorism. Among those arrested in the fracas before the Cathedral was an old revolutionary, a certain Alexis Bogolyubov. The police caught him and he was sentenced to a term of imprisonment. For a technical breach of prison regulations, the Governor of St Petersburg, General Trepov, ordered Bogolyubov to be flogged. The revolutionaries demanded vengeance for this illegal act. Before they could move, a young woman, Vera Zasulich, took the law into her own hands and shot Trepov, wounding him seriously. She made no attempt to escape. Hers had been an act of pure vengeance. She knew neither the prisoner nor the General.

All the radical intelligentsia of St Petersburg watched the Zasulich trial. It was held under the new legal reforms of the 1860s (complete with jury). And the jury acquitted Zasulich, to the rejoicing, not only of the revolutionaries but even of high officials and army officers – a most significant criterion of the isolation of the régime. Zasulich herself was spirited away to Switzerland.

Thereafter, all the trials of those resisting the authorities were to be conducted by military courts. This, together with the increasing use of administrative exile, was symptomatic of the increasing repression exercised by the police in their desperate struggle against the revolutionaries. Terrorist attempts grew bolder and bolder. In broad daylight, in the streets of St Petersburg, the famous Stepniak struck down General Mezentsov, a hero of the Crimea and Chief of the Third Section. In Kharkov, Goldenberg killed Prince Kropotkin, cousin to the anarchist and Governor-General of the city.

Logically, the policy of assassination must end in regicide; and in April 1879 a certain Alexander Solovyov did in fact fire at the Tsar. But he escaped unhurt. 'Land and Liberty' had given Solovyov the gun he used but not the authority to use it. This act, in conjunction with the growing terrorist wave, brought to a head the conflict inside the party on the whole policy of terrorism. In June 1879, Plekhanov, who was defeated on this issue, formed with one or two sympathizers a purely Populist group under the name 'Black Partition', i.e. 'all land to the peasants'. 'Land and Liberty' was dissolved and a new organization took its place – 'People's Freedom'. Its programme described the aim of terrorism as 'lifting ... the revolutionary spirit of the people and its faith in the success of the task ... the Party must take upon itself the start of the overthrow and not wait for the moment when the people will be able to act without its aid,' 'History is terribly slow, it must be pushed forward,' said Zhelyabov, one of the terrorists.

The Executive Committee of the new party concentrated its energy on the assassination of the Tsar. He was formally condemned to death in the late summer of 1879. But not until two years later did the terrorists actually carry out the execution.

'People's Freedom' organized at least seven attempts on the Tsar's life before succeeding. The plotters favoured dynamite and high explosives, made popular in Russia through the war with Turkey. The fatal blow came in March 1881. The Tsar was returning from watching two Guards battalions at their manoeuvres. As his carriage and its mounted escort passed along the Catherine Quay, the first bomb was thrown. But it killed only one of the Cossack escorts and a butcher's boy. The Tsar stepped out of his carriage to have a closer look at the attacker, pinned against the quay, and also the two victims. This took perhaps five minutes. The Tsar then turned to re-enter his carriage and the second bomb was thrown. This was the fatal explosion. He died an hour or so later.

'The nightmare that had weighed down on Young Russia for ten years had vanished,' was the sentiment of Vera Figner, a veteran terrorist, when she heard of the Tsar's death. But despite all the *éclat* of success, it was not to terrorism that the future belonged but to the secessionist, Plekhanov, and his pupil – Lenin.

CHAPTER 12

Dress Rehearsal for Revolution

ONE assassination – even that of a Tsar – does not make a revolution. The *coup* unleashed none of the anticipated upheaval. History refused to be pushed. The masses in town and country refused to budge. The terrorists' appeal to the new Tsar, Alexander III, to call a national assembly and institute a free social order, in return for which they, the terrorists, would suspend their activity, also fell on deaf ears. At both levels – as threat or as trigger – terrorism had failed and was, inevitably, discredited. It did not of course cease to exist. The indomitable Vera Figner, who lived until 1942, a nonagenarian in Moscow, maintained the tradition, and at intervals during the next three decades ministers and lesser dignitaries were sporadically picked off by the 'People's Freedom' and their successors. But terrorism, as a force that would produce the revolution, had died at the moment of its greatest apparent success.

Revolution, on the other hand, although it too shared in the terrorist failure, soon recovered from the general radical discredit. By the end of the century scattered Marxist groups were at work in the Empire of all the Russias, to say nothing of the intensive study of Marxist thought in progress at Geneva and elsewhere. A race for the future developed. To whom did the twentieth century belong? To the Tsar or to the revolutionaries? In this contest it soon became clear that the revolutionaries had a far greater sense of reality than their opponents.

All through the nineteenth century one of the most remarkable features of the autocracy was its loss of zeal, its virtual inability to react, its blind adherence to the *status quo*. It had lost its verve and its nerve. The very fact that so many of its most determined opponents came from the young members of highly placed families showed how it was being undermined from within. It is also significant that at this time there was enunciated almost the only *rationale* of absolutism produced in

Russian history. Hitherto the autocracy had managed without having to produce much of a *raison d'être*. It could take itself for granted. But now no more.

Its spokesman was Pobedonostzev, who proclaimed a pure theory of repression, decrying freedom of the Press, constitutionalism, rationalism, the goodness of human nature, and exalting the virtues of orthodoxy, the family, obedience, and governmental coercion. Come hell and high water, the autocracy must be preserved as the safeguard of the Russian future – this was Pobedonostzev's constant refrain to Alexander III, whom he served as confidant and adviser.

The first victim of the new course was a semi-constitutional project drawn up by Loris-Melikov, the Minister of the Interior. It had actually been approved by Alexander II on the eve of his assassination. At the very beginning of the new reign both the project and the Minister were hurled into limbo. This anticipated the future. In many ways Alexander III harked back to the dark days of Nicholas I, with their slogan of 'autocracy, orthodoxy, and nationalism'. But it was now incomparably more difficult. A much more self-conscious revolutionary movement had developed in the meantime; and the taste for freedom and local self-government and a relaxed censorship, however circumscribed, was not lightly lost.

To achieve the same result, Alexander's régime had therefore to be far more repressive than that of Nicholas. The first move was an attempt to restore the power of the landed nobility and the bureaucracy in the countryside. A new official post was created – that of land captain. He was picked as far as possible from the ranks of the local nobility and exercised judicial and administrative functions. The régime did away with the Justices of the Peace, except in the larger towns, their functions being assimilated to those of the land captains. The latter in fact wielded wellnigh omnipotent power in the villages.

This was in 1889. The next year saw the revision of the constitution of the *zemstva* so as to guarantee paramount power to the noble element and correspondingly diminish the peasant element. In a similar spirit, two years later, the Government raised the property qualification for voters to the organs of municipal self-government. In order to bolster the waning economic power of the nobility, a Nobles' Bank had been founded in 1885. This made loans to landowners at far more favourable terms than did the Peasant Bank. (The latter made loans to the peasants to enable them to purchase land over and above their allotment.)

The same motives inspired an attack on the independence of the judiciary and the inviolability of the courts. Emergency powers, military tribunals, a tighter hold over the judges, granting powers to administrative officials – these were typical of the methods used. Refusing Jews the right to practise at the bar was another device.

The Government gave equal attention to the spread of ideas at every level – from the village elementary school to the university, from the newspaper to the workers' free library and reading-room. Each sign of intellectual life that threatened to diverge from or contradict the official patriotic line fell under closer and closer supervision. University autonomy was undermined, the Press censorship tightened up, village schools subjected to the Holy Synod, and the provision of reading-rooms subordinated to the whim of provincial governors.

The drive for conformity went even further and took on even more sinister forms. In the Ukraine, White Russia, Lithuania, and Poland, the teaching of the vernacular was forbidden in schools and the use of Russian enforced. In the Baltic provinces, Livonia, Estonia, and Courland, the Government exercised similar discrimination against the German element. Forced conversions from Lutheranism to Greek Orthodoxy were common. The Russian nonconformist sects, such as the Doukhobors, the Molokany, and the Stundists, had to suffer religious oppression, deportation, and imprisonment.

It was probably the Russian Jews who now had to undergo the worst torments. Concentrated in the western and southern provinces of the Empire, in the so-called Pale of Settlement, and living at a miserably low level of existence as factory proletariat or in the interstices of the economy – they were petty traders, innkeepers, coachmen, tailors, money-lenders, cobblers, cabinet-makers – the Jews formed an easy target for Tsarism in its death-throes. Anti-semitism had been endemic in Russia for generations. But it had only constituted a 'problem' since the partition of Poland at the end of the eighteenth century. Even so, it had not yet been used as an instrument of government policy.

What happened after the assassination of Alexander II was therefore unprecedented. The Government introduced the *canaille* into the class struggle in a way paralleled only by twentieth-century Fascism. A Jewish girl, Hessia Helfmann, had been one of the Tsar's assassins. This was the pretext for a wave of Government-inspired pogroms that brought terror, death, and rape to more than a hundred Jewish localities in south Russia in the spring and summer of 1881. The outrages were at their worst in Kirovo and Kiev. One-third of the Jews of Russia must die, one-third emigrate, and one-third assimilate, said Pobedonostzev. At the end of the year and early in 1882 the movement spread to Warsaw and Podolia.

The so-called 'May Laws' followed. They forbade Jews to settle in rural districts, even within the Jewish Pale of Settlement. Within the next decade, repressive and discriminatory laws fell thick and fast on the hapless Jews – for example, the quota system in schools and universities, exclusion from the bar, and, for Jewish doctors, exclusion from employment with public authorities, the loss of franchise rights in *zemstvo* and municipality. The process reached its climax in 1891 and 1892 when, without warning and in bitter winter, the Government evicted many thousands of Jewish artisans from Moscow and cleared the Jews from a wide belt of territory on the western frontiers. The Jews were forcibly 'resettled' in the ghettoes of the interior.

This destructive policy was relieved by a certain relaxation of the pressure on the peasants. The redemption payments, dating from the Emancipation, were scaled down. The poll-tax also went. The Government also made it easier, partly through the Peasant Bank and through the removal of legal restrictions, for village communes to rent or buy additional land.

On the other hand, the commune was preserved, with all its power over the individual member. Until 1903, for example, a peasant still had to obtain a passport to leave his village and work outside it. In other respects, the commune was artificially preserved from the free play of economic forces by prohibiting the sale or mortgage of its land except with special administrative permission, and then such sales could be made only to peasant purchasers. But the attempt to 'freeze' the commune could not overcome the division into rich and poor that was steadily making itself more and more obvious among the peasants. It was even more irrelevant to the basic problem – the inadequacy of peasant landholdings. This it was that made such an important contribution to the upheaval of 1905 and to the final cataclysm of 1917.

The contrast between agriculture and industry was startling. The first stagnated, the second flourished. Under the inspiration of Sergei Witte, an outstanding negotiator and organizer, all branches of Russian industry and production showed a remarkable upsurge in the nineties.

Witte, born in 1849, began his career as a student of mathematics at the University of Tiflis. Thence he became a clerk in the ticket-office of the Odessa Railway Company. From this curious rung he climbed higher and higher in railway administration and logistics, earning special praise for his handling of troop movements in the Russo-Turkish war of 1878. He passed into the Civil Service in 1891 as Director of the Railway Department in the Ministry of Finance. The next year he became in rapid succession Minister of Transport and Minister of Finance.

His superiority in the Cabinet was so manifest that he was for all practical purposes the Tsar's chief minister. He made the industrial development of Russia his overriding aim. Not only that: his Russo-Persian bank and his Trans-Siberian Railroad served as agents of Russian expansion and penetration into the Middle and Far East.

A few figures are enough to show the astonishing upsurge in production as Russia prepared to enter the twentieth century. In the last decade of the old century the smelting of pig-iron increased in Russia by 190 per cent (the equivalent rate for Germany was seventy-two, for the United States fifty, for England eighteen). By 1900 Russia had moved up from seventh to fourth place in world production. The same applies to the production of iron, coal, oil, and cotton. In every case the rate of growth in Russia far outstripped that of any other nation. Part of this fantastic rate derives, of course, from the abnormally low starting-point. Even so, the picture remains striking.

In its general lines of development, the same characteristics prevailed as in the earlier period, but to an intensified degree – the emphasis on heavy industry, particularly railroad and rolling-stock construction, the concentration of industry in large units of production, the great extent of governmental financial participation, and the influx of foreign capital.

It is remarkable, for example, that by 1902, 49.8 per cent of the working population was engaged in factories employing a thousand or more workers. The number of such factories had increased between 1879 and 1902 to 123 per cent. The growth in productivity was concentrated in the larger units.

This enormous investment was not financed by private capital raised within the country. It derived almost wholly from Russian government funds and from foreign capitalists and entrepreneurs. The Government, as ever, took a particularly important part in railroad construction. Not only did it issue special loans for this purpose, but it also guaranteed the interest on private railroad loans. In addition, the Treasury was authorized to buy from private railroad companies both their more important and their less profitable lines. The bulk of the investment capital came from direct and indirect taxes and from the deposits in the State savings banks. There developed a special form of State capitalism with special interests in railroads, and through the railroads, in all the ramifications of heavy industry also.

The State's participation in industry went even further. It founded credit institutions. At the State-owned mines in the Urals, Siberia, and the Altai it processed ores for industrial purposes. It enjoyed monopolistic rights in the sale of vodka to the populace. A balanced budget, a highly protectionist tariff, and government contracts were other State-inspired concomitants of the policy of industrialization. At one time, in the early 1900s, the State's involvement in the economic process even went so far as to lead the authorities to suppose that they could organize, through Zubatov, the head of the Moscow Security Police, a number of police unions that would help to contain the growing labour movement.

The encouragement of foreign capital was Witte's special preserve. It was his chief motive for bringing Russia on to the gold standard in 1896 and making the paper rouble convertible into gold. This was the background to the astonishing influx of foreign capital in the 1890s, especially after the ratification of the Franco-Russian alliance in 1894. France and Belgium together accounted for the bulk of foreign investments at this period, followed by Germany and Great Britain. The absolute figures are obscure, but it is estimated that foreign investors supplied something more than one-third of all company capital during 1890, a proportion which, by 1900, had risen to nearly one-half. In some industries, such as mining, it actually exceeded the participation of Russian capital. And of course it was not only capital that came from abroad: Russian industry could also profit from the most advanced technical innovations of the West, could use its technicians, its managers, its manifold specialists. Russia jumped several stages in industrial development in a few decades.

Eight main industrial areas developed: the Moscow region, predominantly textile in character; St Petersburg – metal-processing and machine-building; the Polish region – textiles, coal, chemicals; Krivoi Rog and Donetz – coal, iron-ore, basic chemicals; the Ural mining region; the Baku petroleum region; the sugar-beet area of the south-west; and the manganese region of Transcaucasia.

The human accompaniment to all this was a sudden, gigantic growth in the size of the urban working class. It more than doubled between 1865 and 1890, increasing not only faster than the population as a whole, but also at a faster rate than the urban population. In the next decade the rate accelerated even more. By the beginning of the twentieth century, the total was touching on two and a quarter million. By now, also, a hereditary working class existed. It was no longer formed of peasants displaced from the countryside and often returning to their villages for the harvest: it was a true factory proletariat.

Its dwelling quarters in the new industrial centres were fantastically overcrowded, unhygienic, and squalid. The workers might be herded into factory barracks, fed in factory canteens, and forced to take their wages in the form of goods and foodstuffs.

At a time when the workers in the West, in Britain, France, and Germany, were slowly being integrated into the national life, their worst abuses remedied, and the labour movement acknowledged, the workers in Russia enjoyed no State intervention and were truly beyond the pale. They had no right to strike and no right to form trade unions. Such labour legislation as there was had mainly a perfunctory character and gave no protection against exploitation. This was the immediate background to the growth of Marxism and to the workers' role in the revolutions of 1905 and 1917 – even though it was overshadowed in both by that of the peasants.

THE OPPOSITION

The structure of the Russian economy was such that organized long-term opposition on a class basis, as distinct from sporadic strikes or sporadic upheavals in the countryside, did not exist. Both the big bourgeoisie with its attendant corps of specialists and managers, and the nobility, despite its increasing loss of influence, were too closely linked to the autocracy and the bureaucracy to play a radical role. How many revolutionists does Chekhov's work show? Barely one – and that in a vast panorama of Russian life, embracing every type. The main danger seemed to come from the *zemstva* with their recurrent demand for a *zemsky sobor*, a general assembly of the land, and from the terrorists with their policy of assassination. But if the assassination of the Tsar had no effect, then it was unlikely that that of any other personality would seriously matter. All seemed set fair.

This was a mistaken assessment, however comprehensible. It would be far truer to say, as Engels did in 1883: 'Russia is the France of the last century.' In particular, autocratic complacency took no account of the work of a young man born in 1856, the scion of a middling estate-owner – George Valentinovitch Plekhanov. To Plekhanov, whose early work reared, said Lenin, 'a whole generation of Russian Marxists', belongs the merit of first seriously applying the Marxist analysis to Russian conditions. Plekhanov was by no means the first Russian to interest himself in Marxism. As early as 1860 Professor Babst of the University of Moscow had delivered a lecture on the first part of the *Critique of Political Economy*. Professor Sieber of Kiev was another academic expounder of Marx. In the early 1860s Bakunin had translated into Russian the *Communist Manifesto*; and in 1872 appeared a Russian translation of the first part of *Capital*, the first translation from the original German into any foreign language. Moreover, both Marx and Engels were in regular contact with Russian exiles such as Lavrov. There were assuredly any number of channels through which Marxist influence could flow into Russia. As early as the mid 1870s there were small and short-lived Marxist groups calling for the overthrow of the autocracy.

But all these swallows did not make a summer until Plekhanov appeared. Plekhanov had begun his political career as a member of 'Land and Liberty', but had broken with the Party on the issue of terrorism. Plekhanov condemned as pointless the policy of individual assassination. At the end of 1879, on account of his revolutionary activities, he fled abroad to live in Switzerland. Here, together with other ex-Populists – Vera Zasulich, Paul Axelrod, and Leo Deutsch – he founded a group known as the 'Emancipation of Labour'. This was in 1883. It was a tiny group numerically, so tiny that when the members were out boating on the Lake of Geneva, Plekhanov, one of the few witty Marxists, once quipped: 'Be careful: if this boat sinks, it's the end of Russian Marxism.'

As it happened, for better or for worse, the boat did not sink. But why should it have carried a cargo of Marxists? Why not one of the other revolutionary doctrines of the West? This is no easy question to answer. There are, of course, obvious affinities between Marxism and Populism, such as the belief in violent revolution and the belief in the masses as its agent. There is a strong messianic element common to both. Moreover, even though the industrial proletariat formed a minute proportion of the population of Russia, it was at least growing fast, whereas the peasantry, Populist experience showed, were incapable of responding to any revolutionary appeal. Also, there already existed Marxist Parties in Germany and France.

The point is further complicated by the fact that Marx himself had taken up a somewhat ambiguous attitude towards Populism. He had been prepared to admit that, in certain circumstances, Russia, by way of the peasant commune, might pass directly from a bourgeois to a socialist form of society. But this point is really unimportant. What matters is that Plekhanov's Marxist-inspired attack on Populism succeeded, or at least laid the basis for success.

The struggle was fought out in terms of Russian economic development. More particularly, it was concerned with the role of capitalism in Russia. Was it an alien, unnatural growth, condemned to a premature death? Or was it to be the inevitable next stage in Russia's economic development? Populist economists such as Vorontzov and Danielson argued that capitalism in Russia could make only slow progress because Russia had neither capital nor a bourgeoisie, that the internal consumption level was too low to absorb its products, that it was too weak to conquer markets abroad in the normal imperialist fashion, and that it was kept alive only through massive State subsidies and bounties of all sorts. In the end, therefore, neither having markets at home nor being able to acquire them abroad, it would infallibly collapse.

The Populists concluded from this analysis that Russia could well represent a special case and escape the capitalist phase of economic development. By fortifying the commune and its collective virtues, Russia could evolve its own form of socialism without undergoing the preliminary capitalist phase of exploitation and wage slavery.

To all this, in the polemic-filled eighties and nineties, Plekhanov opposed two basic arguments. First, in such works as *Socialism and the Political Struggle, Our Differences*, and *In Defence of Materialism*, he reiterated time and time again that 'the present as much as the future belongs to capitalism'; and, he added, 'we suffer not only from the development of capitalism, but also from the scarcity of that development.'

Second, Plekhanov argued, Russian capitalism was a progressive movement in that its further expansion would require the overthrow of the autocracy and at the same time lead to the ripening of the industrial proletariat; and the latter in their turn would inaugurate the socialist revolution. At the Foundation Congress of the Second International in Paris in 1889 Plekhanov summed up his creed in its application to Russia: 'The revolutionary movement in Russia can triumph only as the revolutionary movement of the workers. There is no other way out for us, and cannot be. ...' To the Western European socialist leaders this general thesis was a commonplace. But in Russia it was brilliantly original – although it was to be considerably falsified by events.

In practical political terms, Plekhanov's analysis required that the emphasis be shifted from the peasantry to the workers as the principal revolutionary force. The workers would not of themselves be able to carry out the bourgeois revolution. But their skill and energy in leading all democratic forces would determine the interval between the bourgeois and the proletarian revolution.

During the years while this controversy with the Populists was being fought out, small Marxist groups of workers and intellectuals inside Russia were spontaneously coming into being. What redeemed them from possibly ephemeral significance was the participation of Vladimir Ilyich Ulyanov, known to history as Lenin.

He was born in Simbirsk (now Ulyanovsk) on the Middle Volga in 1870. His father was a progressive-minded schoolteacher who rose to become director of elementary schools for the province of Simbirsk. The mother had also been a schoolmistress. All their five surviving children became revolutionaries. The eldest son, Alexander, was executed in 1887 for his share in a terrorist plot to assassinate Alexander III.

At the local high school Lenin had a brilliant record. His headmaster – even in Russia it was a small world – was the father of Alexander Kerensky, the leader of the Provisional Government overthrown by Lenin in February 1917. 'Highly gifted, industrious, and punctual,' ran the verdict of Kerensky *père*. Lenin then entered the local University of Kazan. A few months later the authorities sent him down because of his part in a student demonstration. Not until 1891 was he permitted as an external student of the University of St Petersburg to qualify as a lawyer. He emerged with a first-class diploma and top results in every subject. A little while later he returned to Samara, on the Middle Volga, where he practised as a barrister.

It was not a successful bourgeois career that Lenin sought. He had already familiarized himself with Marx's *Capital*, using his dead brother's copy; and in St Petersburg and Samara he was already in contact with Marxist circles and known for his sharp criticism of Populist ideas. In his first work, *The Position of the Peasants in Russia*, he emphasized the disintegrating influence of capitalism in the countryside. In *What the Friends of the People Are and How They Fight Against the Social-Democrats* Lenin carried on the underground struggle against the Populists, displaying a remarkable mastery of Marxist thought, the art of polemical writing, and the arguments of his enemies.

Lenin combined his theoretical work with practical instruction to Marxist study circles, composed mainly of intellectuals but not without a sprinkling of working men. Many of Lenin's audience came through introductions from Krupskaya, a school-teacher and his future wife.

Vladimir Ilyich [she writes] read with the workers from Marx's *Capital*, and explained it to them. The second half of the studies was devoted to the workers' questions about their work and labour conditions. He showed them how their life was linked up with the entire structure of society, and told them in what manner the existing order could be transformed. The combination of theory with practice was the particular feature of Vladimir Ilyich's work in the circles.

In 1895 Lenin also helped to form the St Petersburg Union of Struggle for the Liberation of the Working Class, which was active in the great strikes of the middle and later 1890s in the capital. In the same year Lenin applied for a passport, ostensibly on health grounds, to which colour was given by a recent attack of pneumonia. But it was in actual fact to meet Plekhanov and the other exiles in Geneva. Lenin spent four months in Western Europe – France and Germany, as well as Switzerland. Axelrod noted afterwards: 'I felt that I had before me a man who would be the leader of the Russian revolution. He was not only a cultured Marxist – of them there were many – but he also knew what he wanted to do and how to do it. He had something of the smell of the Russian soil about him. ...' This was precisely what the exiles had been missing during their long years in Switzerland. Now their labours, it seemed, were beginning to bear tangible fruit in the person of the brilliant young man from Simbirsk.

Lenin returned to St Petersburg in the autumn of 1895. He had with him a double-bottomed trunk full of illegal literature printed in Geneva. A month or so later the police picked up his trail, aided probably by spies abroad and an *agent provocateur* planted in the study circles.

The Tsarist police, like the Tsarist censorship, were capricious to a fault. Lesser men than Lenin they broke or executed. But in Lenin's four years of imprisonment and Siberian exile he was able to turn his prison cell and then his peasant hut into a well-stocked library. On these resources he drew for his major historical work *The Development of Capitalism in Russia*. He was also able to translate Webb's *History of British Trade Unionism*, to attack the Populists, the 'economist' Marxists who would water down the revolutionary theory into the fight for economic aims and leave political issues to one side, and also the 'legal' Marxists who limited themselves in the main to economic analysis. Thus, when Lenin was released at the beginning of 1900, the world to which he returned had lost none of its familiarity. He took up the struggle where he had left it four years earlier.

His first thought was to found a paper and a Party. An attempt had been made at Minsk in 1898 to establish a Russian Social-Democratic Party, formed of scattered Marxist groups and the Jewish Socialist Workers' Party, the *Bund*. But barely was the Party founded than the police seized the founders. The coast was still clear for Lenin to overcome the political and doctrinal fragmentation inherent in the conditions in which Russian Marxism had grown up.

Now the two streams came together – that of the active Marxists in Russia and that of the theoreticians in European exile. It was not an easy marriage. Lenin, Martov (a product of the Jewish Marxist circles in Vilna and the Pale), and Potresov, who had access to funds through wealthy relations, made their way to Geneva, one by one. The union was eventually consummated, and in December 1900 the first number of *Iskra* (*The Spark*) appeared. The editorial board consisted of Plekhanov, Axelrod, and Zasulich on the one hand, and Lenin, Martov, and Potresov, on the other.

Iskra was printed in Munich, Leipzig, Stuttgart, and London, in tiny characters on thin cigarette paper, and smuggled into Russia by way of Romania, Prussia, and Austria. (It was as illegal in Germany as it would have been in Russia itself. The capitalist international was no less alert than the socialist.) A network of agents saw to its further distribution in the great industrial centres. One such group existed in Tiflis, where the young Stalin was undergoing his own revolutionary apprenticeship in the workers' circles. The distribution system under Lenin's control was in fact an instrument for the inculcation of ideas, the organizational network of a Party that did not yet exist.

In these early years *Iskra* was an attempt to weld into one the scattered Social-Democratic groups inside Russia. It had to fight for the Marxist solution as interpreted by Lenin, against that of the 'Economists' and the 'legal' Marxists. In 1900 a new enemy had appeared – the Social-Revolutionary Party, heirs to the Populists and soon to become the chief Party of the left, far superior in numbers to the Social-Democrats.[1] But mere numbers were not decisive. Organization and policy were, and it was to these that *Iskra* made an invaluable contribution.

The very first issue contained an article by Lenin closely anticipating the sort of Party he hoped to form:

Not a single class in history has reached power without thrusting forward its political leaders, without advancing leading representatives capable of directing and organizing the movement. We must train people who will dedicate to the revolution, not a spare evening but the whole of their lives....

This was the immediate prelude to the formation of such a Party of revolutionaries. A second congress – so-named in deference to the first, abortive congress held at Minsk in 1898 – opened in Brussels in 1903. Half-way through, the delegates moved to London to escape the attentions of the Belgian police. This was the crucial congress. It saw the birth of Bolshevism as a political organization and a special form of political warfare that had no precedent in the West, and also the split with Menshevism. Present were forty-three delegates with fifty-one votes, representing either Russian underground or *émigré* organizations. The disused flour warehouse in Brussels, swarming with rats and fleas, witnessed some of the stormiest and most emotional debates that even the Russian revolutionary movement had produced. The decisions of the congress were to impress an indelible stamp not only on the immediate struggle against the autocracy, not only on policy in 1917, but also on the type of ruling Party that would emerge after the revolution.

After the singing of 'The International', the congress turned to serious business. What would constitute, for example, the Party programme? This was divided into a minimum and a maximum. The first demanded the overthrow of the autocracy, the erection of a democratic republic on the basis of a constituent assembly elected by free, equal, direct, and universal suffrage, freedom of speech, Press and assembly, equality before the law, and the introduction of the eight-hour day. The maxi-

1. The genesis and policy of the Social-Revolutionaries are discussed on pp. 216–17.

mum programme called for the socialist revolution and the establishment of the dictatorship of the proletariat. It was in connexion with a hypothetical conflict between the unexceptionally democratic demands of the first part and the dictatorship explicitly called for in the second part that Plekhanov made his celebrated declaration: he denied that universal suffrage was 'a fetish', and emphasized that *salus revolutiae [sic] suprema lex*. ... If the salvation of the revolution should demand the temporary limitation of one or the other democratic principle, then it would be a crime to hold back. ...' With an implicit reference to Cromwell, Plekhanov continued:

> If the people, seized by revolutionary enthusiasm, were to elect a good Parliament, then we must try to make of this a Long Parliament. If the elections, on the other hand, go badly, then we must try not to wait two years to dissolve Parliament, but to do so after two weeks if possible.

Acclamations and also some hisses greeted this exposé. But, saving one abstainer, the congress adopted the programme.

The basic conflict emerged in the debate on Paragraph 1 of the Party statutes. How define a Party member? What was his duty? To Lenin a Party member was 'any person who accepts its programme, supports the Party with material means, and personally participates in one of its organizations'. To Martov the criterion was less rigid. He would replace 'personally participate' by 'cooperate personally and regularly under the direction of one of the organizations'. This trifling verbal distinction did in fact conceal and involve far-ranging divergencies on an interrelated hierarchy of values. Was revolution an inevitable and spontaneous process, or did it require the intervention of a conscious force? More specifically, must the Russian socialist revolution wait on the achievement of a bourgeois revolution, or could it be the by-product, as it were, of the bourgeois revolution? Finally, and as a consequence, must a socialist Party in Russia be organized so as to be a minority, bringing to the workers 'from outside' their revolutionary consciousness, or must it seek to envelop as many of the workers as possible? In other words, was the Party to be a disciplined body of professional revolutionaries, or was it to be a mass Party somewhat similar to those that had developed in the West?

Lenin's determined insistence on the need for professional revolutionaries, as distinct from Party members who might be nothing more than well-wishers or sympathizers, did not spring forth fully formed, like Minerva from Jupiter's head. It was drawn from the experience of the Russian revolutionary movement extending over the best part of a century. Tsarism, to a very large extent, imprinted its own image on its enemies. A centralized authoritarian autocracy called into being a precisely similar revolutionary movement. Already Pestel, the Decembrist, had understood the need for conspiratorial groups. But Tkachev, a follower of the French revolutionary Blanqui, and one of the leaders of the People's Freedom Party, was probably Lenin's closest precursor, in his view that revolutionary success depended on the work of an underground corps of professional workers for the cause. Lenin himself once paid tribute to his predecessors:

The magnificent organization that the revolutionaries had in the seventies should serve us all as a model ... no revolutionary tendency, if it seriously thinks of fighting, can dispense with such an organization ... the spontaneous struggle of the proletariat will not become a genuine class struggle until it is led by a strong organization of revolutionaries.

This in itself could not, of course, guarantee success. But in Russia at least experience showed that anything else invariably succumbed to the police. These were the Russian roots of Bolshevism. From the same environment stemmed the need for secrecy, the impossibility of public discussion of policy, and the election in open forum of Party leaders.

Lenin made the same point in his criticism of the 'Economists'' proposals for the organization of workers' factory circles. Their rules demanded, for example, keeping a record of events in the factory, collecting funds regularly, presenting monthly reports on the state of the funds – 'Why', exclaimed Lenin, 'this is a very paradise for the police; for nothing would be easier for them than to penetrate into the ponderous secrecy of a "central fund", confiscate the money and arrest the best members ... only an incorrigible Utopian would want a *wide* organization of workers, with elections, reports, universal suffrage, etc., under the autocracy.' (Italics in the original.)

The counterpart to Lenin's demand for a disciplined corps of professional revolutionaries was the realization that the masses of themselves could only develop what he called a 'trade-union consciousness', i.e. could not rise above the level of immediate economic demands. But this would make a political revolution impossible. Thus Lenin's

theory of Bolshevism amounted to acknowledging that the revolutionary forces had to be re-created and organized outside and even against the 'immediate interests' of the proletariat, whose class-consciousness had been arrested by the system in which they functioned. The Bolshevik doctrine of the predominant role of the Party leadership as the revolutionary vanguard grew out of the new conditions of Western society (the conditions of 'imperialism' and 'monopoly capitalism') rather than out of the personality or psychology of the Russian Marxists.[1]

What Lenin did was to assimilate to Russian conditions both the theory and practice of Marxism. As he himself said: 'Different in England from France, different in France from Germany, different from Germany in Russia.'

Not all this was by any means evident in 1903. After the congress Lenin himself sought for reconciliation with his defeated opponents. But it was not to be. Lenin lost the debate on the Party Status by twenty-eight votes to twenty-two. On the other hand, the Leninist group was victorious in the elections to the editorial board of *Iskra* and to the Central Committee. This majority, often very slender, earned for it the description *Bolshevik*, from the Russian word for majority, and for Lenin's opponents the description *Menshevik*, from the word for minority. But there was still no hard and fast division, and in 1904 the Central Committee fell to the Mensheviks, Lenin

1. Herbert Marcuse, 'Dialectic and Logic' in *Continuity and Change in Russian and Soviet Thought*, ed. E. J. Simmons, Harvard, 1955, pp. 348–9.

resigned from *Iskra* and published his own journal, *Vperyod* (*Forward*). In 1905, however, the division went considerably further when the Bolsheviks and Mensheviks held separate congresses, in London and Geneva respectively.

Trotsky belonged consistently to neither faction. But it was he, on the very morrow of the split, who brilliantly foresaw one consequence of Lenin's theory of organization. 'The Party is replaced by the organization of the party, the organization by the central committee, and finally the central committee by the dictator.'

'A SHORT, VICTORIOUS WAR'

After 1878, with its setback to Russian ambition in Europe, there was a natural tendency to give priority to expansion in the Far East. In 1875 there had already been a Russo-Japanese deal whereby Russia abandoned to Japan the Kurile Islands in her possession in return for the Japanese renunciation of all claims to the island of Sakhalin. Quite apart from the general trend eastwards, the Sino-Japanese war of 1894–5, with its dual revelation of Chinese weakness and the emerging power of Japan, and the construction of the Trans-Siberian Railroad pulled Russia into the competitive imperialisms of the Far East. The special area of penetration was Manchuria. A treaty with China in 1896 gave Russia the right to construct a railroad across the northern part of the territory. A later treaty entitled Russia to construct northwards from the military and naval base of Port Arthur and the commercial port of Dalny the South Manchurian Railroad, to link up with the Chinese Eastern Railroad.

The Boxer Rebellion of 1900 gave a further boost to a forward Russian policy by justifying the stationing of troops in Manchuria. But in 1903, the planned evacuation of these troops, forced on Russia by Anglo-Japanese diplomatic pressure, the opposition of the United States, and the reluctance of France any longer to support her ally's Far Eastern advances, failed to take place. Moreover, the Tsar, who had by now largely taken the control of foreign policy into his own hands, sanctioned the infiltration of Russian troops into Northern Korea, and the exploitation of a timber concession near the Yalu River.

In 1903 the tension between Russia and Japan could have been resolved had the former accepted a Japanese proposal for the mutual acknowledgement of each other's respective preponderance in Manchuria and Korea. But the Russian interest in Korea eliminated this possibility. There came instead a sudden Japanese attack on Port Arthur.

The war was not unwelcome in certain Russian governing circles. Apart from its meaning to the country's foreign policy, it would also, Plehve hoped – and Plehve was the Minister of the Interior – be 'a short, victorious war that would stem the tide of revolution'. In the same spirit Plehve instigated a number of pogroms – at Kishinyov on Easter Sunday 1903, and again at Gomel in August–September 1903 – in the hope of 'drowning the revolution in Jewish blood'. Jewish blood was indeed shed. But it did not produce the desired effect. On the contrary, the revolutionary wave swelled to unprecedented size and scope, and all the more so when the war in the Far East turned out to be far from short and far from victorious.

For at least the previous decade the radicalization of all sections of Russian society had been taking on more and more extreme forms. The exact number of strikers is unascertainable. It seems to have varied from 17,000 in 1894 to 87,000 in 1903, with ups and downs in between. But there is no doubt that the general trend was upwards. Although many were organized by the St Petersburg Union for the Liberation of the Working Class, the strikers' aims were primarily economic and not political. A reduction in working hours, a Saturday half-day, higher wages, a reduction in fines – these were the aims of the strike movement. In the main, the Government's answer was repression, the arrest and deportation of the strike leaders. Another method, of course, was the Zubatov movement.[1]

In the countryside, peasant discontent was aggravated by a series of famines – in 1891–2, 1897, 1898, and 1901. But the unviable character of rural life was evident enough without that. Unmanageable tax arrears, agricultural stagnation, rural

1. See p. 200.

ARCTIC

BARENTS

Murmansk

SEA

BALTIC SEA   FINLAND

GERMANY
E. LAT EST
PRUSSIA Riga St Petersburg   Archangel   ARCTIC
Warsaw   Grodno R. Volkhov
Vilna   Novgorod
Lublin Brest Litovsk
Zhitomir Minsk
Mogilev   TRANS- SIBERIAN R.Ob   R. Yenisei
Kiev   Moscow
UKRAINE   Perm
Odessa   Ekaterinburg
R. Dnieper Kharkov   Tobolsk
R. Volga   Ufa Cheliabinsk
BLACK   R. Don   Orenburg   Omsk Tomsk
SEA   R. Novosibirsk
TURKEY   Trans-Caucasia   CASPIAN SEA   R. Irtysh
Kars Tiflis
Aral Sea   L. Balkhash
Baku   Syr Daria
Khiva   Turkestan
Amu Daria Bokhara
PERSIA   Tien Shan
Pamirs

AFGHANISTAN

5. THE RUSSIAN

O C E A N

Bering I. St.

Anadyrski Mts.

Koryaksia Mts.

Cherskogo Mts.

Verkhoyanski Mts.

Vilyuiski Mts.

Kamchatka

CIRCLE

R. Lena

Dzhugdzhur Mts.

SEA of
OKHOTSK

Yakutsk

Stanovoi Mts.

Sakhalin

Bureinski Mts.

Krasnoyarsk

Khabarovsk

Blagoveshchensk

Yablonovy Mts.

Lake Baikal

Nerchinsk

Irkutsk

TRANS-SIBERIAN RLY.

MANCHURIA

Vladivostok

SEA of
JAPAN

M O N G O L I A

J A P A N

0   200   400   600   800   1000
Miles

EST. • ESTHONIA
LAT. • LATVIA
LITH • LITHUANIA

over-population, the paralysing effects of communal tenure, all emphasized that the existing order was doomed. In 1902 peasant uprisings in the provinces of Poltava and Kharkov gave added point to the threat of imminent collapse.

This background of rural violence gave birth to a more radical version of the Populist groups – the Social-Revolutionary Party. It was founded in 1900 as a sort of 'umbrella Party' of the left. Although sharing many Marxist presuppositions, the S.R.s never produced a coherent statement of policy. 'Their programme', it has been said, 'disclosed an effort to link up Populism with Marxism and Henry George.'[1] They had common ground with the Marxists in their insistence on the socialist potentialities of the coming bourgeois revolution. At the same time the S.R.s were proposing a union of all anti-bourgeois forces – intelligentsia, industrial workers, and poor peasantry. Even the Liberals might fit into this all-embracing scheme, said the S.R.s. From the Populists the S.R.s inherited the belief in a specific Russian form of agrarian development, based on the commune. They advocated the nationalization of the land with a view to its distribution on the basis of use.

As regards tactics, the S.R.s combined mass propaganda with reversion to the policy of terror and assassination. A special 'Fighting Organization' carried out a number of notable *coups* in the years 1902–7. The leaders were Gershuni, Boris Savinkov, and Eugene Azef. It was typical of the political *demi-monde* in which terrorism operated that the last-named was simultaneously a police agent, planted in the revolutionaries' ranks by Plehve. It was also typical that Azef, of Jewish origin, supervised, though he did not execute, the assassination of Plehve in 1904. Other casualties were the Minister of the Interior, Sipyagin, the Governor of Ufa, and the Governor-General of Finland.

Lastly – the Liberals. For almost the first time in Russian history liberalism began to develop, as an independent political force. Its main base was the *zemstvo* movement, which now demanded for itself wider administrative powers and the right to nation-wide organization. Their demands gave some colour

1. D. Mitrany, *Marx against the Peasant*, London, 1951, p. 68.

to the current Menshevik thesis that the break between the bourgeoisie and the autocracy could not be long delayed. At the end of 1904 a meeting of *zemstvo* leaders demanded by a considerable majority the abolition of the autocracy and the installation of a parliament with power to issue laws and to control the budget. The *zemstvo* programme also demanded guarantees of personal and civil liberties, equality before the law, universal suffrage, freedom of the Press, and freedom of assembly.

But this unimpeachably democratic and liberal programme made no mention of the far more urgent questions affecting the actual conditions of the workers and peasants. All the same, on this basis, the Union of Liberation, embracing *zemstva*, municipalities, professional men, corporations of the nobility, and also the Social-Revolutionaries, instituted a bold campaign of reform through political banquets, public assemblies, and nation-wide agitation.

'BLOODY SUNDAY'

The sparks came on 22 January 1905 (9 January old style).[1] Port Arthur had fallen a few weeks earlier, the latest and most crushing of a series of Russian defeats in the Far East. At much the same time a strike broke out in St Petersburg at the Putilov engineering works. It rapidly spread to other factories. Here, overnight almost, was a mass workers' movement of unprecedented dimensions. The tension rose to such a pitch that it forced a certain Father Gapon, who led a police labour union, either to take some positive action or to abdicate entirely and leave his members to follow an even more radical path. Gapon, in collaboration with some of the *zemstvo* leaders, intelligentsia, and Social-Revolutionaries, drafted a petition for presentation to the Tsar. It was couched for the most part in plaintive terms – 'we are not considered human beings . . . we are treated like slaves'. But it also contained outspoken political demands: freedom of the Press, religion, assembly; the calling of a con-

1. All dates from now on have been converted to the contemporary Western Calendar.

stituent assembly; equality before the law; labour legislation and the eight-hour day; a reduction in indirect taxes and introduction of a graduated income tax; an amnesty for political prisoners, and an end to the war. One hundred and thirty-five thousand people signed the petition.

And so, on Sunday, 22 January, the petitioners, of whom there were about 150,000, marched under Father Gapon's leadership to the Winter Palace. It was a peaceful and unarmed procession. The marchers intoned hymns, bore icons, and portraits of the Tsar.

But the sound of hymns gave way to the crack of bullets when the crowd reached the Winter Palace. There was no deliberate decision to shoot, it seems. But the authorities refused to receive the petitioners. (The Tsar was not in residence.) Salvo after salvo poured into the terror-stricken marchers. At the end of 'Bloody Sunday' there were perhaps a thousand dead and many more thousands wounded.

This was the spark that set alight the flame of revolution. In all social groups, in all parts of the country, revolt flared up. By the end of January nearly half a million workmen were on strike. The professional intelligentsia joined in. Doctors, lawyers, teachers, professors, engineers, formed unions to press political claims on the Government. The terrorist movement flared up again with the killing of the Tsar's uncle, the Grand Duke Sergey, who had made himself infamous for his cruelty during his Governor-Generalship of Moscow. Industrialists also joined in the clamour for a constitutional régime. A new feature of the popular movement was the formation of an All-Russian Peasants' Union – the first time that a political Party of the peasants had arisen on Russian soil. It soon joined the Union of Unions, led by the Liberal, Paul Milyukov. In many parts of the country a state of anarchy prevailed that all the Tsar's courts martial and repression could not abolish.

The Tsar, meanwhile, was hoping for military victories that would re-establish his shattered prestige. There came, instead, defeat at Mukden in Manchuria, and the destruction of the Russian fleet at Tsushima. This, in conjunction with incipient rebellion, forced the Tsar's hand. In August a law was issued that promised a consultative assembly, known as the Duma, to be formed from an electoral body weighted in favour of high property qualifications and also the peasantry. But this succeeded in splitting off only the right-wing Liberals. To the rest of the Empire it was matter for derision and boycott. To the workers, particularly, the proposed Duma meant nothing. They would have no vote at all.

In the spring the first wave of strikes petered out, and the initiative passed to the middle classes. But in the summer and autumn the workers' movement suddenly revived. Not that the summer had seen any real break in the movement: troops had been called out in Lodz, and in the Black Sea fleet the crew of the battleship *Potëmkin* had mutinied – the subject of one of Eisenstein's most famous films. Also, of course, there were few provinces, and especially the border regions, where the peasants were not plundering the manors, despoiling the landowners' estates, raiding store-houses, burning land registers.

The events of the autumn eclipsed even these. In the second half of September, a printers' strike in St Petersburg touched off, with unexampled rapidity, what was in all but name a general strike. It came so swiftly and spontaneously that even the revolutionaries were taken by surprise – as indeed they had been by most of the events of the year. Barricades sprang up on the streets of Odessa, Kharkov, Ekaterinoslav. The whole life of the country was paralysed. There were further mutinies among the troops. The countryside, too, was ablaze with peasant violence. The climax came, both from the Government's standpoint and that of the revolution, in the middle of October. The first was the constitutional manifesto that Witte virtually extorted from the Tsar, since Witte realized that by now the very existence of the autocracy was at stake. The second was the St Petersburg Soviet. These twin phenomena towered above the chaos and tumult of the year. The difference between them? One came too late, the other too early.

The constitution proclaimed certain fundamental civil liberties, promised to extend the franchise beyond the limits announced in July, and promised also that no law would be promulgated without the approval of the State Duma.

The St Petersburg Soviet, formed of some 500 delegates elected by about 200,000 workers, represented the peak of working-class achievement. It was a spontaneous creation. But it was a lesson in revolution, not the revolution itself. The Soviet, of which Trotsky was at one time the co-chairman, followed on the whole a moderate policy. It supported the general strike, but sought rather to educate the workers in the limitations of the Duma than to organize an immediate armed insurrection. 'A constitution is given, but the autocracy remains,' said Trotsky. The most revolutionary act of the Soviet was to issue an appeal for the non-payment of taxes and the withdrawal of bank deposits.

In the meantime, the October manifesto had to some extent split off the liberal middle class from the workers. The Government, demoralized as it was, could still proceed with the repression of the revolution. The leaders of the St Petersburg Soviet were arrested and an armed uprising in Moscow put down with relative ease. But even at the end of the year, with mutinies in the Sebastopol fleet, and revolts in Siberia, Batum, Kharkov, it might seem that a renewed upsurge was in the making. In actual fact, the autumn had seen the climax. Not for more than a decade would it be overtopped by a second and even more tumultuous upheaval.

CHAPTER 13

Forebodings, 1881–1917

BY the 1880s the great age of Russian realism was over. A Russian La Bruyère might well have said: '*Tout est dit.*' The poet Nekrasov died in 1878, Dostoevsky in 1881, Turgenev in 1883, the dramatist Ostrovsky in 1886, and Goncharov, the creator of Oblomov, in 1891. What could remain to be said? What could still be written that would be more than repetition?

Such pessimism would have been misplaced. The new age called forth new talents and a new relationship to reality. To Gorki, speaking in 1934 at the first Congress of Soviet Writers, 'the main and basic theme of pre-revolutionary literature was the tragedy of a person to whom life seemed cramped, who felt superfluous in society, sought therein a comfortable place, failed to find it and suffered, died, or reconciled himself to a society that was hostile to him, or sank to drunkenness or suicide'.

The onslaught on society through the description of the fate inflicted on its more talented members was indeed a basic theme, both before and after the 1880s. But it ceased to be treated on the same scale; and it ceased to resound with the same note of protest. The isolation of the individual from society came to be matter for acceptance rather than matter for attack. But this was no joyful acceptance. It was combined with a fierce rejection of the values of society and a very strong undercurrent of foreboding. It was the note that Tolstoy struck in a diary entry in 1881: 'An economic revolution not only may but must come. It is extraordinary that it has not come already.' The famous words that Chekhov put into the mouth of Baron Tuzenbach in *Three Sisters* are even more prophetic:

The time has come, an avalanche is moving down on us, a mighty, wholesome·storm is brewing, which is approaching, is already near, and soon will sweep away from our society its idleness, indifference, prejudice against work, and foul *ennui*. I shall work, and in some twenty-five or thirty years everyone will work too.

Round about the turn of the century such writers as Freud, Sorel, and Nietzsche, had already given voice to a sentiment of collapse, that a world was dying, that the old order had had its day, that forces of unreason and violence were struggling to the surface. This foreboding and anticipation of the future was sensed in Russia perhaps more deeply than anywhere else – if only, perhaps, because violence and the régime's unvarying policy of repression were so patent a feature of the national life.

Thus to those unable to share in the hope of revolution the future was bleak indeed. And never more so than in the period of intensified reaction, of pogroms and oppression, that followed the assassination of Alexander II. It was then that Tolstoy, having renounced literature, became the moral conscience of the nation, and grew to such stature that although the Government might prohibit the publication of his works, it dared not silence the man himself.

'Who is mad – they or I?' Tolstoy asked this question after a family discussion in which he had heard conventional pro-Governmental views put forward on the topics of the day. The question would not obsess him much longer. His positive doctrine was thin enough – the gospel of universal love, undogmatic Christianity, sexual abstinence, the renunciation of tobacco and alcohol, non-resistance to evil. But his disintegrating criticism of society and its values, his corrosive and derisive scepticism, made him into an anarchist more anarchic and a nihilist more nihilistic than any whom Russia had yet produced or would produce in the future.

'What is science?' he asked. Had it done anything important or unique when it determined the weight of Saturn's satellites? What was universal suffrage? A means for the prisoners to elect their own gaolers. Had industrialism raised the standard of living? Then look at the slums and doss-houses of Moscow. In his powerful work *The Restoration of Hell* (1903), he derided division of labour as a device for turning men into machines, book printing as a medium for communicating 'all the nasty and stupid things that are done and written in the world', philanthropy as plundering by the hundredweight and returning by the ounce, socialism as fostering class enmity in the name of the supreme organization of man's life – and as for reform, 'it taught people that though themselves bad they can reform bad people'. No iconoclast ever made such a *tabula rasa* of the hopes and beliefs cherished by the men of his time.

After such denunciation, the voice of Chekhov must inevitably sound muted and anticlimatic. But Chekhov too had his own specific contribution to make to the revolt against society that marked the Russian *fin de siècle*. It was couched primarily in personal terms. To politics Chekhov was more or less indifferent. He was explicitly apolitical.

Great writers and artists [he wrote in defence of Zola's intervention in the Dreyfus case] must take part in politics only in so far as it is necessary to put a defence against politics. There are enough prosecutors and gendarmes already, without adding to the number.

Chekhov, however, in his own way, and despite himself, was also a prosecutor; and the indictment that he drew up against Russian society is redolent of Gogol's *Dead Souls*. But it is expressed with a mitigating touch of sympathetic humour where Gogol had been satiric. To Chekhov Russia was a country where men, kindly men and men of goodwill, were isolated both from one another and from society. It was a chaotic, frustrating world of alienated individuals, victims of their own sentiment of inferiority *vis-à-vis* an impossibly imposing environment. It was a world where the individual was powerless, and reduced to finding his *raison d'être* in the expression of his own feelings. It was a world where nobody was happy.

Chekhov did indeed have faith in a better future. His vision was not irretrievably and inconsolably bleak, as was Tolstoy's. But the situation was so desperate that half-measures were a mere palliative; and he could not conceive of revolution. In the end, all that was left to comfort Chehov was a belief, as he himself put it, 'in individual personalities, scattered here and there throughout Russia, whether they be intellectuals or peasants'. How substantial could this hope be, in the face of the overwhelming preponderance of the trivial, the humdrum, the shoddy, and the pretentious?

The Russian *fin de siècle* began in earnest after the failure of the 1905 revolution. The specifically Russian form of the decadent movement emphasized the impossibility of communication between one human being and another, which is, for example, the leitmotive of Andreyev's writings. There was also the turning away from public themes to private pursuits – the glorification of sex, the resurrection of myth and legend, the invocation of death, the exploration of the emotions. These last were the favoured matter of the symbolist poets, the closest Russian equivalent to the Western European exponents of art for art's sake.

The literary ferment of these years was accompanied by a cultural renaissance in the arts. These are the years that saw the first flowering of the Moscow Arts Theatre of Stanislavsky and Nemirovich-Danchenko, the work of Meyerhold at St Petersburg, the 'Diaghilev period' of the Russian ballet, the first works of Stravinsky, the development of some of the most advanced ideas in socialist and religious thought. In Gorki the literary movement produced the first consciously proletarian novelist. For the first time in their history St Petersburg and Moscow became intellectual and artistic centres comparable to any in the West.

But all stood under the sign of collapse. In the decades preceding the First World War there was an outburst of chiliast, apocalyptic, and eschatological thought and sentiment. This had always been a feature of Russian social criticism. Bakunin's anarchism, for example, has been seen as the direct continuation of chiliast attitudes in the modern world; and the messianic conception of Moscow as the Third Rome had been preserved in many of the dissident sects, less subject to the Westernizing influence of St Petersburg. It even emerged in the Petrashevsky circle, where French Utopian socialism took on at times an unusual note.

A life of wealth and bliss, to cover this beggar's earth with palaces and fruits and to adorn it with fruits – that is our aim [wrote one of the members]. Here in our country will we begin this revolution, and the whole world will bring it to fulfilment. Soon the whole of mankind will be freed from unbearable suffering.

The nearer the revolution came, and particularly after 1905, the more acute and pervasive this feeling grew. It can be found in Merezhkovsky, the theoretician of symbolism, and also in the works of the religious philosophers, Rosanov and Solovyov. It even survived the Bolshevik Revolution of 1917 and inspired the 'Scythian' movement of Ivanov-Razumnik and the peasant-poet Essenin. The 'Scythians' saw the Revolution as the outburst of a newborn, unlimited, maximalist spirit that would sweep away the corrupt old world, as Christianity had once burst asunder the decadent Roman Empire. In the 'International' Ivanov-Razumnik could hear the words 'Peace on earth, goodwill to men'. This cast of thought reached its climax in Blok's poem, 'The Twelve', of 1918. Blok, probably the greatest of the symbolist poets and a sympathizer with the left Social-Revolutionaries, who were themselves allied for a time to the Bolsheviks, depicts twelve riotous, rowdy Red Guards marching through the snowy streets of St Petersburg, shouting, shooting, cursing, blaspheming. They are led by a man bearing the red flag – Jesus Christ.

In his essay of 1918, 'The Intelligentsia and Revolution', Blok compared the revolution to a mighty phenomenon of nature. 'We Russians,' he exclaimed, 'are living through an epoch which has had few things equal to it in grandeur.' The aim is –

to remake everything. To build everything anew, so that our lying, dirty, boring, monstrous life becomes a just and clean, a joyous and beautiful life. . . . The range of the Russian Revolution aimed at embracing the world. A true revolution cannot wish for less. . . . 'The peace and the brotherhood of the people' – that is the symbol under which the Russian Revolution is taking place. It is of this that its torrent is roaring. This is the music which those with ears must hear.

This dithyrambic outburst was not typical of all the intelligentsia. Blok himself died disillusioned in 1921. But it forms an apt prelude to the last decade of the autocracy. Blok's cry: 'We are Scythians, we are Asiatics, with slit and hungry eyes' was tantamount to a rejection of the whole St Petersburg tradition.

CHAPTER 14

From Revolution to War

THE revolution of 1905 did not end with the arrest of the St Petersburg Soviet or the crushing of the Moscow armed uprising. In the early part of 1906 the peasant movement raged as intensively as ever. To take one instance: in one district of the province of Voronezh the peasants razed to the ground all the landowners' houses and farms. Mutinies in the army and navy were almost as frequent in 1906 as in the preceding year. But it is clear in retrospect that the high-point had come, and gone.

'*La Révolution est morte, vive la Révolution,*' exclaimed Trotsky. The words had a challenging ring, but also a somewhat hollow sound. One of the Tsar's achievements in 1905–6 was to inflict a crushing blow on the workers' movement, to cripple and maim it for the best part of a decade – as an active force, that is. The declining number of strikers tells its own story – in round figures, 1905 – three million; 1906 – one million; 1907 – 400,000; 1908 – 174,000; 1909 – 64,000. No wonder Lenin's wife, Krupskaya, could lament in 1909: 'We have no people at all.' The active membership of the underground revolutionary groups declined from some 100,000 in 1905 to some 10,000 in 1910.

Internationally, however, as Lenin pointed out, the abortive revolution had given 'rise to a movement throughout the whole of Asia. The revolutions in Turkey (1908), Persia (1909), and China (1911) prove that the mighty uprising of 1905 left deep traces, and that its influence expressed in the forward movement of *hundreds and hundreds* of millions of people is ineradicable'.

And yet, and yet ... by 1906 the revolutionary élite lay in Tsarist gaols, in Siberian deportation, or in European exile. Lenin was back in Finland. A year earlier he had been zealously translating into Russian a manual on street-fighting tactics. Trotsky, clad in drab prison garb, was on trial for his life. He would shortly be on his way with fourteen others to serve a sentence of life deportation in Siberia.

The year of revolution had shown the astonishing resilience of the régime. It could yield ground on every front, be bitterly opposed by almost every stratum of society, be crippled by a general strike, see outbreaks of mutiny in the army and navy, lose a disastrous war in the Far East, be discredited internationally – and still survive. And not only survive; it could also go over to the offensive. As early as October 1905, hundreds of pogroms in the towns and hamlets of the Jewish Pale brought terror and destruction to a wide area. In later years this method of government came increasingly to be used. In 1911, for example, the Government even went so far as to accuse a Kiev Jew, Mendel Beilis, of murdering a Gentile child for ritual reasons. The case, a *cause célèbre* in its day throughout the world, ended two years later with the acquittal of Beilis.

For the moment, however, a more immediate task was to confront the Duma and the new political Parties that sprang into being. This did not prove over-difficult. The State Council, the upper chamber, consisted half of Crown nominees and half of elected members – representative in both cases of the wealthier commercial and professional classes. The State Duma, a purely elective lower body, was formed on the basis of the law of August 1905 as subsequently amended in December 1905 and February 1906. The electors were divided into six *curiae*, depending on whether they were landowners, town-dwellers, peasants, or workers. Each *curia* elected electors in the proportion of 1 to 2,000 (landowners), 1 to 7,000 (town-dwellers), 1 to 30,000 (peasants), and 1 to 90,000 (workers). This rigged assembly was further hampered by the narrowness of its competence. The Duma could exercise only a very limited control over the State finances. There was virtually no responsibility to the Duma on the part of the Council of Ministers; they were responsible only to the Tsar. Lastly, no law could become effective without receiving the approval of both houses and the sanction of the Tsar.

Given these handicaps, it is a tribute to the optimism of the voters that this first experiment in Russian constitutionalism yet gave birth to a number of Parties and groupings, and that by 10 May 1906 some 450 deputies were present at the opening of the Duma.[1]

There were about twenty-six political groupings and some twelve national groups. The Constitutional Democrats – Cadets, for short – who represented liberal and *zemstvo* opinion had 170–80 seats. Their more moderate wing, the Octobrists – i.e. those who took their stand on the basis of the October manifesto – had 30–40. The national groups – Polish, Ukrainian, Latvian, etc. – with some 60 seats were radical and nationalist. A Labour group, mainly peasants but politically uncommitted, had a combined membership of about 100. There were another 100 peasant members of no definite affiliation, and 18 Social-Democrats, mainly Mensheviks. (The official policy of the Social-Democrats was to boycott the elections, but this was not everywhere observed.)

It must have been a strange opening session in the throne-room of the Winter Palace! To one side of the Tsar stood Senators and the Imperial entourage; to the other the newly elected deputies, many in peasant blouses, workers' overalls, and rough boots. So charged with hostility was the mood that Stolypin, Minister of the Interior, whispered to his ministerial colleague, Kokovtsov, that he feared a bombing attempt. The Dowager-Empress noted 'a strange, incomprehensible hatred' on the faces of the deputies – as well she might.

The storm burst at the Duma's first session. Speaker after speaker brought forth flaming demands – an amnesty for political prisoners, the abolition of the death penalty, the resignation of the Government, the elimination of the upper house, ministerial responsibility, the confiscation of the large estates, the right to strike, equality before the law, the reformation of the whole tax system, and a democratic electoral system.

Each demand in itself was tantamount to a declaration of war. Taken together, they destroyed any possibility of co-operation between the Government and the Duma. In ten

1. The total number of deputies was theoretically 524; but many of those from the more remote parts of the Empire did not arrive in St Petersburg until after the Duma had been dissolved.

weeks it was all over. On 22 July (9 July old style), a Sunday, large bodies of troops surrounded the Tauride Palace, where the Duma met, and simultaneously the Tsar decreed its dissolution on the grounds that it had exceeded its competence, shown itself incapable of efficient work, and addressed an illegal appeal to the people. The Tsar gave no date for new elections, as he was legally obliged to do, but merely proclaimed that the second Duma would convene in February 1907. Two hundred left-wing and Liberal members thereupon made their way to Vyborg in near-by Finland. Here they appealed to the public to refuse to pay taxes and to refuse army service. But this form of resistance was futile and brought no public response.

The second Duma followed a similar path to the first and was dispersed after a three-month session on trumped-up charges against a Social-Democratic deputy. He was accused of plotting the assassination of the Tsar and of inciting the troops to mutiny. The third Duma ran its full course. But by then the electoral law had been changed to favour the landed gentry and wealthier town-dwellers; and the fourth found itself obliged to yield to the Provisional Government of March 1917. Altogether it was not a happy history. The Duma suffered from what has been called

... the dilemma of attaining complex, specifically Western objectives in an illiberal, under-developed society. Were the only alternatives those of conciliating the illiberal government or changing it by illiberal means...? On each occasion the range of plausible Liberal answers was narrowly circumscribed by the same environment, the same Russia.[1]

To return to the first Duma – the dissolution was precipitated by the land issue, more precisely, by the need to alleviate the peasants' land-hunger. This was not fortuitous. Both the Government and the Parties of the centre and the left – apart, of course, from the Social-Revolutionaries who had their own built-in peasant policy – were beginning to concentrate their efforts on fortifying the peasants as a bulwark of the State, or,

1. George Fischer, *Russian Liberalism*, Harvard, 1958, p. 203.

alternatively, on mobilizing the discontent of the peasantry as a positively anti-governmental force rather than allowing such discontent to dissipate itself in futile, sporadic, and short-lived *jacqueries*.

Ever since the late 1890s some sort of agrarian reform had been in the air. Arrears of redemption payments were mounting, and the size of the individual allotments was becoming tinier and tinier as a result of population growth. The issue soon began to centre on individual as against communal tenure. Would Russia become a country of peasant proprietors? Certain steps presaged this development. The power of the commune was weakened in 1903 when, in many provinces, it was relieved of joint responsibility for the State obligations of its members. Then, in 1905, a decree cancelled all further redemption payments after the peasants should have paid half the sums due for 1906. This was a radical move.

In the first Duma, Professor Herzenstein, a Cadet member and Moscow agricultural specialist, proposed that the State expropriate the larger landowners on payment of fair compensation. The land would then be distributed to the peasants in return for a form of payment. But this smacked too much of emancipation with strings, *à la* 1861, to be of much use.

It was from the Government itself that some of the most revolutionary measures relating to Russian land tenure finally emanated. Stolypin was their progenitor. He was born the son of a landowner, graduated from St Petersburg University, and, like his father before him, took up a career in the bureaucracy. By 1905, when Stolypin was forty-four, he had risen to become governor of the province of Saratov. In that year of revolution he made himself notorious for his cruelty to peasants and revolutionaries. His later actions, especially his ruthless treatment of Jews and Poles, further justified this reputation. The field courts martial and extraordinary powers given to governor-generals were responsible, between September 1906 and May 1907, for the execution of more than 1,100 people.

Repression, however, was only one half of Stolypin's policy. He combined with it two basic ideas regarding the peasant. Both marked a notable advance on previous governmental reasoning. First, Stolypin argued, the peasants could be used to defeat the revolution; second, that to do this the Government must free them from the commune and permit them to acquire private property.

The ultimate idea was to introduce almost completely free trade into the buying and selling of land. The more able peasants would then emerge as small landed proprietors with a strong stake in the existing order; and they would hold in check the less able, who were destined to dwindle into a landless rural proletariat. In Stolypin's words, 'the Government relied not on the feeble and the drunk, but on the solid and strong'.

This policy came into effect in 1906, when peasants were allowed to obtain passports on the same terms as anyone else and when the land captains lost their power over the peasants. It continued until 1911, when the final land-settlement act was passed. The intervening legislation was almost as complex as the Emancipation Act of 1861. But essentially what it did was to entitle every householder in a commune where the land was periodically redistributed to turn his share of arable land into his inalienable, individual property. The commune could not oppose this. Where the communes were based on hereditary tenure, the Government simply decreed that individual ownership was now in force. It was a draconic measure.

Many other rural institutions went into limbo together with communal tenure. Communally owned pasture and grazing land was divided up; the traditional scattered strips of land were consolidated into one farmstead; finally, joint family ownership was, as a rule, abolished by acknowledging the family elder as the owner of the household's allotment.

Altogether some twelve million peasant households were involved and 320 million acres. The reforms got off to an excellent start in 1907 with some 50,000 householders leaving the commune; the number soared the next year to half a million, and in 1909 to nearly 580,000. The machinery came to an end at the beginning of 1916 because those peasants in uniform were naturally unable to take part in the negotiations over their property rights. By then some two million householders had received legal title to their new lands and started out on their own. Many, of course, later sold their land to the more able peasants, and in this process the enlarged Peasant Bank played a growing and important role. The net result was growing social stratification in the village. This was accompanied by modernized methods of farming, increased production by the purchase and rent of extra land, and expanding output in the more profitable branches of agriculture. Internal colonization was another means of attack on the peasant problem. Between 1906 and 1915 about three and a half million people emigrated to Siberia, usually with Government assistance.

But it was all too little and too late. The problem was only partly one of redistributing the land. The average peasant holding in Russia by 1905 compared very favourably with that of its French or German counterpart. But agricultural technique was so low that its yield in crops was no more than between one-half and one-third. Second, Stolypin's reforms took no account of a growth in rural population that amounted to some one-and-a-half million per year. Thus, even if he had thrown into the balance the 140 million-odd acres that were still in the hands of the nobility – which was politically out of the question – he could still not solve the problem. In fact, he actually exacerbated it by increasing the number of landless peasantry and grinding them down both absolutely and relatively. What was needed was the raising of agricultural productivity and accelerated industrial development and to this the reforms made only an insignificant contribution in relation to the size of the problem.

Lenin, and the revolutionaries generally, no less than Stolypin, could draw the lessons of the peasant upheavals of 1905. The land issue came up at the fourth Party Congress at Stockholm in 1906. In all respects this congress showed a greater sense of reality than that voiced in Plekhanov's declaration that the Russian revolution would triumph as a proletarian revolution or not at all. At the congress a further step was taken in applying revolutionary doctrine to Russian conditions. After 1905 and the formation of the Peasant Union in Moscow, it was no longer possible to shrug off the peasant movement as an inchoate *jacquerie*. But this is not to say that unanimity prevailed. The Mensheviks argued in favour of expropriating the land and municipalizing it for the benefit of local peasant bodies. Lenin advocated the nationalization of the land, or, failing that, encouraging the peasants in any direct seizure of the land. Stalin, the sole Bolshevik representative in the Caucasus delegation and just beginning to win his spurs in the oil country of Baku, put in a plea for the redistribution of the land from the landowners to the peasants.

Much of this was, of course, the very antithesis of the basic tenets of Marxism, which could not but be opposed to the creation of a propertied petty-bourgeoisie stratum, backward and essentially reactionary. But Lenin had the honesty to admit that in 1905 he had underestimated 'the breadth and depth of the democratic or rather bourgeois-democratic movement among the peasantry'. On the other hand, he reinterpreted the situation in such terms as to assert that what the Russian peasantry were carrying out was not the struggle against large-scale farming, but the struggle of the dispossessed against feudalism. Thus they were carrying the class struggle into the countryside and clearing the way for the emergence of capitalist farming, which would produce its own proletariat: '... in this instance,' Lenin declared, 'we want to support small property, not against capitalism but against feudalism.'

But this was an extremely dangerous game to play. What if the decision was taken out of the Marxists' hands and 'the other side' organized the struggle against feudalism? The Stockholm conference was held before Stolypin's policy was announced. When this came into the open, there would necessarily be unpleasant implications for the Marxist cause. Of course, a landless rural proletariat would be created with a burden of discontent that could easily be exploited. As against this, however, a bourgeois peasant class would come into existence, clinging to its new status and property, and constituting a far stronger barrier to revolution than a declining rural gentry, riddled with debts, incompetent, and often only too happy to dispose of its property at a reasonable price. By 1911, however, Lenin could discern the failure of Stolypin's attempt 'to pour new wine into the old bottles, to reshape the autocracy into a bourgeois monarchy'. He drew comfort from the fact that this failure 'is the failure of Tsarism on this last road – *the last conceivable road* for Tsarism'. (Italics in the original.)

Here was some consolation for the fact that from 1906 to 1912 the revolutionary movement was undoubtedly in the doldrums. Not the least of its worries were financial – a problem only partially overcome by the use of 'expropriation' tactics. These were the notorious bank raids and hold-ups waged against State and private institutions for the purpose of seizing cash, and at times weapons also. Since this policy gave rise to recurrent scandals, and since the raids were primarily a Bolshevik prerogative, they were an added source of dissension between Bolsheviks and Mensheviks.

But for all the factional disputes, all the controversy between Bolsheviks, Mensheviks, 'conciliators', 'liquidators', and Trotsky (who played a lone hand), the fact is that the decade preceding the revolution brought forth virtually all the leaders of the future Bolshevik State. The fight against the autocracy in these years, both open and underground, gave birth to the first generation of Bolshevik statesmen and political personalities. In foreign affairs, there was the contrasting pair of Chicherin and Litvinov – the one a scion of an aristocratic and intellectual family, the other of an intellectual family from the Jewish Pale; in foreign trade, Krassin, an engineer employed in Russia by Siemens and the vast German electrical combine, A.E.G.; at home, Bubnov, Commissar for Education; Tomsky, trade-union leader; Voroshilov, Commissar for War; Rykov, vice-chairman of the Council of Commissars; planners such as Kamenev, and fiery, unscrupulous politicians such as Zinoviev, first president of the Communist International – not to mention, of course, Trotsky and Stalin. This galaxy of talent, intellect, and ability was all born in the fierce and sometimes fratricidal strife of the pre-revolutionary years.

Contrast this with the men of Tsardom – it is Hyperion to a satyr. The only two men of stature were Witte and Stolypin. The first was ignominiously dismissed when he had done his duty to the dynasty, and the second was assassinated, perhaps

by one of the very people whom he had planted in the terrorists'
ranks. Otherwise the picture of the Tsar's Ministers and con-
fidants shows a succession of mediocrities, barely distinguishable one from the other. The human factor swung all in favour
of the Marxists.

ACCOMMODATION WITH THE WESTERN POWERS

After the defeat of 1905, the story of Russia's foreign relations
can largely be told in terms of their accommodation to the
policies of Britain and France. As a corollary, the quasi-tradi-
tional Russo-German alliance, or *entente*, gradually gave way
to hostility and rivalry.

The first move was a drawing together of victor and vanquished in the Far East. The threat of American economic
intervention in northern China, an area where both Russia and
Japan had staked out their own claims, hastened this *rapprochement*. In 1907 a Russo-Japanese convention confirmed
the division of Manchuria into a northern, Russian sphere of
influence and a southern, Japanese sphere. Russia implied by
this an abandonment of its more ambitious plans with the Far
East and an affirmation of the *status quo*. Later revisions of
this agreement gave it more bite. Japan annexed Korea, and
Russia more or less created the client, puppet state of Outer
Mongolia.

This relatively pacific policy in the Far East enabled Russia to
retire gracefully from the Far Eastern imbroglio and to return
to Europe. It was a development encouraged by Great Britain
and France, the former linked to Japan, the latter to Russia,
and both linked by the Entente Cordiale. The same thing had
happened, but in a reverse direction, after the defeat of 1878.
This choice – either Europe or the East – was always open to
Russia. The choice depended on the extent of the counter-
pressure. Now, baulked in the East, Russia returned to Europe.

The Anglo-Russian convention of 1907 accelerated this process. In three disputed areas it defined the relations of the two
countries. It was a classical specimen of the diplomacy of the
period. In Tibet, both Powers acknowledged Chinese sov-

ereignty and undertook to refrain from intervention; in Afghanistan, Russia recognized an English sphere of influence; Persia was carved up, the north with its oil wells falling to the Russians, the south, of more use to a naval power, falling to the British. A central, buffer zone was left free. Russia accepted British hegemony in the Persian Gulf area in return for support in the region of the Dardanelles.

The exact purport of this latter aspect of the agreement was never clear. It therefore became quite a source of tension between Russia and Britain. But relations were none the less improved. Even so, it was certainly not this Convention that brought Russia into the orbit of France and Britain, but rather the conflict in the Balkans and in Turkey. The first brought Russia up against Austria-Hungary and the second brought Russia up against Germany. In an area seething with nationalist discontent, the rivalries of the great Powers added a particularly heady ingredient. To Russia it slowly became clear that the aim of free passage for its warships through the Straits could not be achieved without war.

The first crisis came in 1908–9 when Austria-Hungary unexpectedly annexed Bosnia and Herzegovina, two Slav principalities in the Balkans. Serbia reacted vigorously, encouraged by some Russian support. But German support for Austria was far stronger, and a virtual ultimatum to Russia forced it to drop its backing for Serbia. The Russians were in a way happy to withdraw. Stolypin realized only too well that the army was in no condition to fight a European war.

But this was the prelude to further Russian involvement in the Balkans. What happened in 1908 was a *reculer pour mieux sauter*. Sazonov replaced Izvolski as Foreign Minister, and over the next two years created a Balkan League of Serbia, Bulgaria, Greece, and Montenegro that would serve as a means of strengthening Russia in the Balkan Peninsula. At the end of 1912 the League attacked Turkey and was eminently successful. The Greeks were in Salonika, the Serbs on the Adriatic, and the Bulgarians at the gates of Constantinople. In the spring of 1913 the Sultan agreed to give up almost all Turkey's European possessions.

But when it came to dividing the spoils of war, the members of the League fell out. Bulgaria attacked Serbia, who fought alongside Greece and Romania and, later, Turkey. The Bulgars were soon defeated and lost their share. Serbia then undertook to conquer Albania, but this met with an Austrian ultimatum based on the determination to keep Serbia, and thereby Russia also, out of the Adriatic.

The overall effect of these wars was, most importantly, to create further hostility between Russia and Austria-Hungary and to add to both great Powers their client states. Serbia and Montenegro fell to Russia, who also enjoyed the support of Romania. Bulgaria turned to Austria and Germany.

The Balkans were not the only area where Russia faced the Central Powers. There was also Turkey. It was a principle of Russian policy in all the Balkan crises that no Power but Turkey should control the Straits. It was thus that German intervention in Turkish affairs aroused considerable Russian nervousness. First came the project for a railroad from Constantinople to Baghdad, the concession for which was to be financed by a German group. The Russians knew only too well to what use a concession in an economically weaker country might be put. Russian opposition in this case was bought off by German acknowledgement of Russian supremacy in northern Persia.

Russia scored a further slight success in countering German intervention when the Young Turks set about modernizing the régime after the revolution of 1908. The Turks not only engaged the services of a number of English and French military specialists, but also appointed a mission headed by the German general, Liman von Sanders, to modernize the army. Russian protests, however, backed up by French and, to a lesser extent, by British, forced the Turks to clip von Sanders's powers. Enough rivalry remained to range Russia more and more firmly on the side of the Western Powers. To the Franco-Russian military convention of the 1890s was added in 1912 a naval convention; and on the eve of the war a vast loan was floated in Paris for the construction of railways that would facilitate Russian mobilization.

Even so, in 1914, Russian adhesion to the Allied cause was no foregone conclusion. Both then and until the February revolution of 1917 there were strong pro-German trends centring on the German-born Tsarina, on Witte, on Sukhomlinov, the Minister of War (1909–15), on Durnovo, Minister of the Interior in 1906, and on certain German-orientated industrial and banking circles. But the pull to war proved stronger. It was also one of the most fateful decisions in all Russian history.

CHAPTER 15

Towards the Abyss

WHEN the Russian State entered the war it did so in an incomparably ramshackle, dilapidated, and inefficient condition. But this did not mean it was powerless. On the contrary, in one way the very fact that it was so inefficient, that power was so decentralized, that the bureaucracy was so incompetent, acted as a source of strength, for it gave to the State a quality of resilience that a more efficient and centralized régime might have lacked.

This resilience had shown itself more than once in Russian history. In the seventeenth century, during the Time of Troubles, the combined effects of civil war, foreign intervention, and economic collapse had still left the State intact. Napoleon's invasion of 1812 was no more successful. Even 1905, which would have toppled many a régime, did no more than shake the autocracy. The German invasion of 1941 is but another example of the amazing resilience of the Russian State. So long as the centre remained standing, almost any sacrifice could be endured, any loss undergone, any disaster overcome.

But resilience and the ability to absorb punishment were not enough in the conditions of twentieth-century warfare – all the more so in the face of the unsolved social problems that beset Russia. Stolypin's policy did, it is true, inaugurate an era of relative peace on the agrarian front. In the years before the war the number of peasant riots annually decreased. But Stolypin had asked for twenty years of peace – and this he would not have. In any case it is not easy to see how his policy could in the long term have contributed a lasting solution. On the other hand, in these very years, the workers' movement was reviving after the defeat of 1905. The number of strikes rose from 222 in 1910 to 466 (1911), to 2,032 (1912), 2,404 (1913), and to 4,098 (January–July 1914).

The dynasty had long lost any prospect of acting as a rallying-point. It had never regained the prestige lost on Bloody

Sunday. The tercentenary of its accession, celebrated in 1913, evoked no answering echo amongst the public. Furthermore, the growing influence of Rasputin, the licentious, hypnotically gifted monk, rendered the court more and more odious.

Even so, as in all the other belligerent countries there was something of a *union sacrée* at the beginning of the war. The war was all things to all men. The ruling classes, as in Germany, could see in war the opportunity of diverting the labour movement into less dangerous paths. The Russian Liberals could console themselves with the hope that alliance with the Western democracies might lead to some relaxation of the autocracy. And as for Lenin – 'a war between Austria and Russia would be a very helpful thing for the revolution,' he said in 1913, 'but,' he added regretfully, 'it is not likely that Franz Josef and Nikolasha will give us that pleasure.'

The first Russian offensive culminated in the defeat at Tannenberg in August 1914. The campaign was undertaken in response to urgent French pleas for a manoeuvre that would take some of the pressure off Paris. And save Paris it did. The East Prussian campaign forced the Germans to divert eastwards two corps needed for the crucial first battle of the Marne. But the Russian losses were 170,000; a few weeks later the Russians suffered another disaster in East Prussia. The next year, on the southern front in Galicia, there came further disasters, with casualties reaching unprecedented totals. By the end of the first ten months of war they have been estimated at 3,800,000. In 1915 also, Poland, Lithuania, and Courland all fell to the Central Powers. In 1916 an offensive against Austria-Hungary dealt the Dual Monarchy an irreparable and overwhelming blow. It also saved the Italian army, materially helped the Allies during the battles of Verdun and the Somme, and brought Romania into the war on the allied side. But again Russian casualties and prisoners ran into millions. By now, the end of 1916, in a welter of corruption, incompetence, abysmal military leadership, and unimaginable human suffering, the Russian army had all but lost its capacity to fight.

Collapse at the rear matched disarray at the front. The mood of August 1914 was ephemeral. Very soon an unbridgeable chasm was opening up between the Government and the people. In fact, who or what *was* the government? It changed so quickly, no one could be quite sure. In the first two years of war, four prime ministers, three foreign ministers, three defence ministers, and six ministers of the interior came and went.

The railway system in the western provinces and Poland proved to be inadequate; again, through the closing of the Baltic and the Black Sea ports, Russia was virtually cut off from its allies. Only the port of Archangel remained open, and this was icebound for about half the year. Furthermore, it had only a limited railway link with the interior. Entry through Vladivostok entailed a journey half-way round the world. The low level of technical and economic development produced an army suffering a paralysing shortage of equipment and trained personnel. Many soldiers often had no weapons at all; they were expected to arm themselves from the discarded rifles of the killed and wounded. Shells had to be rationed to the artillery batteries. Hospital and medical services, in the hands of volunteers from the union of *zemstva* and town councils, were so thinly spread that they had no practical value. The call-up operated irrationally, amounting by 1917 to some fifteen million – about thirty-seven per cent of the male population of working age. Chaos was piled on chaos through the influx of millions of refugees, mainly Jews and Poles, from Poland when the western provinces were placed under martial law and their population moved out of the fighting zone. As early as 1915 it was becoming difficult, according to the reports of provincial governors, to call up reservists and new classes of conscripts. Inflation, recourse to paper money (the convertibility of the rouble was abandoned in July 1914), rising prices, food shortages, and a fall in real wages produced an increasing ordeal for the mass of the population. This applied more to the towns than to the villages, which were more self-supporting. A rapidly mounting wave of strikes gave voice and vent not only to economic but also to political demands. 'Down with the Tsar' – this was the ominous cry beginning to be heard.

This went hand in hand with an atmosphere that Trotsky has described with characteristic verve:

Enormous fortunes arose out of the bloody foam. The lack of bread and fuel in the capital did not prevent the court jeweller, Fabergé, from boasting that he had never before done such flourishing business. Lady-in-Waiting Vyrubova says that in no other seasons were such gowns to be seen as in the winter of 1915–16, and never were so many diamonds purchased ... everybody splashed about in the bloody mud – bankers, heads of the commissariat, industrialists, ballerinas of the Tsar and the Grand-Dukes, Orthodox prelates, ladies-in-waiting, liberal deputies, generals in the front and rear, radical lawers, illustrious mandarins of both sexes, innumerable nephews, and more particularly nieces.[1]

At the end of 1916 and the beginning of 1917, almost every voice spoke in tones of an imminent upheaval. General Krimov told a Duma delegation: 'The spirit of the army is such that the news of a *coup d'état* would be welcomed with joy. A revolution is imminent and we at the front feel it to be so.' The peasants were saying: 'When ten or fifteen generals are on the gallows we shall begin to win.' A Police Department report noted 'a marked increase in hostile feelings among the peasants not only against the Government but also against all other social groups. . . .' The same report stated:

The proletariat of the capital is on the verge of despair ... the mass of industrial workers are quite ready to let themselves go to the wildest excesses of a hunger riot. . . . The prohibition of all labour meetings ... the closing of trade unions, the prosecution of men taking an active part in the sick benefit funds, the suspension of labour newspapers, and so on, make the labour masses, led by the more advanced and already revolutionary-minded elements, assume an openly hostile attitude towards the Government and protest with all the means at their disposal against the continuation of the war.[2]

In a word, the war had utterly destroyed any confidence that still remained between the Government and the people.

The revolution began on 23 February 1917 (8 March old style). This would be the revolution that had been talked over

1. Leon Trotsky, *History of the Russian Revolution*, English translation, London, 1934, pp. 46–7.
2. See M. T. Florinsky, *The End of the Russian Empire*, Yale, 1931, pp. 232, 205, 175–7.

and fought for during the best part of a century. Many of those who had done the fighting and the talking, the plotting and the propaganda, who had printed the illegal leaflets, served their term in Siberia, and run the gauntlet of the secret police, would not of course have recognized their progeny in November 1917. But that is the way of history.

It began in a small way, spontaneously, almost, one might say, unpolitically. It was confined at first to Petrograd – St Petersburg's wartime name. There were strikes, housewives' demonstrations, mutinies among the troops and police – a collapse of all authority. The movement took the revolutionaries by surprise as much as anyone else. Public buildings were taken over against very little opposition, prisoners released, and police stations and barracks captured. The army refused to open fire on the demonstrators.

Side by side with the slow, elemental, mass movement went the 'official' revolution. This was compounded of the aspirations of the Russian Liberals, certain court circles, and British diplomacy, anxious to rid the country of an ineffectual government and to promote the war more effectively. On 12 March (27 February old style) the Premier, the senile Prince Golitsin, informed the Duma that the Tsar had decided on its prorogation. But the deputies refused to disperse and formed ten of their colleagues into a Provisional Government dominated by the Liberals and including also three right-wing members and one leftist, Alexander Kerensky, as Minister of Justice. On 15 March (2 March old style) the Tsar abdicated in favour of his brother, the Grand-duke Michael. But the Grand-duke refused the throne, after due consideration. For the first time in three centuries, Russia was without a Tsar.

More significant was the absence of a centre of power. The Provisional Government could not fill the gap. It was more a testimony to the patent inability of the Imperial Cabinet to carry on in the face of a disastrous war and a demoralized home front, rather than a harbinger of the new world of the masses.

Into this vacuum of power flowed irresistibly, inevitably, the Petrograd Soviet of Workers' and Soldiers' Deputies. This was a rebirth of the Soviet of 1905. Its members were elected by their workmates in barracks, factories, workshops, and public enterprises. It was not long before similar soviets were formed in Moscow and the provincial towns, and also in the country-side. In some places they even controlled food distribution. They undoubtedly enjoyed incomparably more support than the official Government. 'The Provisional Government', said one minister, 'possesses no real power and its orders are executed only in so far as this is permitted by the Soviet of Workers' and Soldiers' Deputies, which holds in its hands the most important element of actual power, such as troops, railroads, postal and telegraph service.' The Soviet's 'Order No. 1' enjoined all soldiers to obey only the Soviet and to safeguard their arms lest they fall into counter-revolutionary hands.

But the Petrograd Soviet made no real use of its power and even supported the Government. There was, it is true, a certain identity of policy between the two bodies. The Government proclaimed an amnesty for political prisoners, abolished discriminatory legislation, inaugurated the eight-hour day, restored the constitution of Finland, promised the Poles independence, undertook to arrange the election of a constituent assembly which would consider the peasant question – with all this the Soviet was in agreement. But as yet it did not go beyond this point, and hence no real conflict developed with the Provisional Government.

The first cleavage came in April when the Minister of War, Guchkov, had to resign. Milyukov, Minister of Foreign Affairs, soon followed. In an incautious note to Russia's allies he had ventured to state that Russia stood by its obligations, i.e. pursued the same policy that had already taken the country into the abyss of revolution. It was thus that the Provisional Government moved more to the left. It had to, merely in order to survive. Six socialist ministers joined with the Cadets – Kerensky, who was associated with the Social-Revolutionaries, became Minister for War; Tseretelli, a Menshevik, Minister of Posts and Telegraphs; Chernov, a Social Revolutionary, Minister for Agriculture; and Skobelev, at one time an associate of Trotsky, Minister of Labour. Prince Lvov remained Premier.

But this leftward shift of power was dwarfed by events in the countryside. Here a genuine and immense agrarian revolution was in progress. The peasants were at last fulfilling their age-long yearning. They were simply expropriating the large estates. Every month the total rose. It was only 17 in March 1917. In April it jumped to 204, in May to 259, in June to 577, and in July to 1,122. By March 1919 virtually all the usable land was in peasant hands. The peasants were not socialists, of course; but it was this elemental movement that was indispensable to the victory of Bolshevism. It was this movement, as much as any other factor, that led to the disintegration of the Russian armies.

'ALL POWER TO THE SOVIETS'

The second stage in the revolution began with Lenin's arrival at the Finland Station in Petrograd on 2 April 1917. From the first Lenin had been in no doubt that the imperialist war must be turned into a civil war. But he had no idea that the trans-formation was so imminent. At late as January 1917 he was telling his audience at a Zürich meeting (Lenin spent the bulk of the war, from September 1914 to March 1917, in Switzer-land), that 'we of the older generation may not live to see the decisive battles of this coming revolution'. And when the first news of the formation of the Provisional Government reached Zürich, Lenin, like the other revolutionaries, did not anticipate all the possibilities that were opening up. None the less, he was all agog to return to Russia. When he did so, it was under the aegis of the German Government. This contrasts ironically with Lenin's view that 'the German proletariat is the most trust-worthy, the most reliable ally of the Russian and the world proletarian revolution'. In reality, the contrary was true.

From 1915 onwards the Germans, like the Japanese in 1904–5, had been fishing in the waters of the Russian left-wing Parties. They promoted revolutionary propaganda in Russia as an act of political warfare. Through Stockholm, in the main, funds flowed into Russia and were used, it may be conjectured, for the publication of pamphlets and the financial support of left-wing groups. At the end of September 1917, on the eve of the October revolution, Kühlmann, the German Secretary of

State, summed up these activities with pardonable self-congratulation:

> The military operations on the Eastern front ... were seconded by intensive undermining activities inside Russia on the part of the Foreign Ministry. Our first interest in these activities was to further nationalist and separatist endeavours as far as possible, and to give strong support to the revolutionary elements. We have now been engaged in these activities for some time, and in complete agreement with the Political Section of the General Staff in Berlin. . . . Our work together has shown tangible results. The Bolshevik movement could never have attained the scale or the influence which it has today without our continual support. There is every indication that the movement will continue to grow, and the same is true also of the Finnish and Ukrainian independence movements.[1]

But this is to anticipate. In March 1917, as one item in this policy, the return of Lenin and a group of other Bolsheviks and revolutionaries was decided on, for the purpose of facilitating further disintegration in Russia. (The Germans little suspected that the boot would soon be on the other foot!) Thirty-two people in all left Switzerland – nineteen Bolsheviks, six Bundists, three Mensheviks, and four miscellaneous. The most important were Lenin, Krupskaya, and Zinoviev.

The journey to Petrograd took Lenin and his party through Germany, Sweden, and Finland. Tumultuous crowds greeted Lenin in Petrograd, the *Marseillaise* thundered forth from a thousand voices, a searchlight played over the faces of the throng, Bolshevik posters and slogans decorated the platform walls and station buildings. A curious and significant encounter then took place in the waiting-room, normally set aside for the Tsar's personal use. Chkeidze, the Menshevik president of the Petrograd Soviet, welcomed Lenin with the assertion that the principal task of the workers was now to defend the revolution, and for this unity was the requisite of the day. But Lenin's reply spoke of the Russian revolution as the harbinger of world revolution. It was not something that had come to a full stop. Its greatest hour was still to come.

1. *Germany and the Revolution in Russia*, edited by Z. A. B. Zeman, Document No. 71, Oxford, 1958.

This preliminary brush was but a foretaste of the thunderbolt that Lenin was preparing to hurl into the Bolsheviks' ranks. After a triumphal procession with speeches at every street corner, Lenin, in an armoured car, eventually arrived at Bolshevik headquarters, the sumptuous house of Kshesinskaya, prima ballerina and one-time mistress of the Tsar. Here there was a reception, snacks, more speeches. It was late at night before Lenin proclaimed the policy that he had been slowly hatching both before and during the return from Zürich. The following day, at the Tauride Palace, he presented this policy to a congress of Bolsheviks in the famous 'April Theses'. He demanded the overthrow of capitalism as the only way to end the war; no further support must be given to the Provisional Government; the power of the soviets must be built up, and the power of the Bolsheviks inside the soviets; there must be no parliamentary republic; the land and the banks must be nationalized; the soviets must take production and distribution into their own hands; a new International must be founded to replace the defunct Second International.

This programme met instant opposition. Only Alexandra Kollontai, the future Bolshevik Ambassador to Sweden, supported Lenin. Of the others, Molotov stood closest to him. But to the majority he was a 'Bakunin', 'a madman', a dealer in 'abstractions'. Yet the key to the Theses is clear enough – that the bourgeois-democratic phase of the revolution was concluded and that what must now be prepared was the transition to the socialist phase, which would be incorporated in government by the soviets. All that Lenin left open was the timing of the socialist revolution. For the moment the slogan was – 'All power to the Soviets'.

After the first upheaval, the new line was adopted at the Petrograd Party conference, and then by the 150 delegates to the All-Russian Party conference. Kamenev, who had been released with Stalin from Siberian exile at the time of the February revolution, was the heart of such opposition as there was. But it never amounted to more than a handful of votes. Thus, by the end of April 1917, the Bolsheviks had committed themselves to opposing any collaboration with the Provisional

Government and to transferring power to the soviets. This was not yet the call for an immediate revolutionary transformation. That would come later.

In the meantime, Kerensky helped to dig his own grave. To some extent, of course, he was the prisoner of the Allies in the matter of war aims and the secret treaties. After America's entry into the war in March 1917, he tried to secure President Wilson's intervention, through the journalist Lincoln Steffens, in favour of the abrogation or at least revision of the treaties. He, Kerensky, would then be able to consolidate the Provisional Government and to pursue the war on terms that might make more appeal to the demoralized Russian army and people. But Wilson denied official knowledge of the treaties; and Kerensky's similar pleas to the British and French Governments were also shrugged off.

None the less, Kerensky went ahead with the preparation of an offensive on the south-western front in Galicia. This was in fulfilment of a Russian pledge to the Allies. Some early successes were secured; but this was largely because the Austrian armies were almost as war-weary and mutinous as the Russians. When the Germans rushed up reinforcements, the offensive turned into a disorderly rout.

In Petrograd, meanwhile, the failure of the offensive sparked off an open insurrection against Kerensky. Demonstrators, mutineers, deserters, unemployed workers thronged the streets. But the movement had no clear political aims. The Bolsheviks were at first indifferent. Later they tried to lead it to success. But in four days it petered out. All the same, it gave Kerensky occasion to suppress the Bolsheviks as defeatists and agents of the Germans. *Pravda* was banned. Lenin and Zinoviev fled to Finland. Kamenev was arrested. Trotsky (who had returned from New York at the beginning of May), Kollontai, and Lunacharsky were arrested shortly afterwards. Kerensky himself, after a prolonged ministerial crisis, became a sort of dictator, dedicated to saving the country from Bolshevism.

The Bolsheviks were now weaker, both in the capital and the provinces, than at any other time in 1917. They owed their rehabilitation to the failure of the insurrection attempted by

General Kornilov, the newly appointed Commander-in-Chief. Politically naïve, Kornilov became a front for right-wing influences to which certain Allied elements were sympathetic. This was a foretaste of the intervention to come. Kornilov tried to march on Petrograd. But his movement disintegrated long before it reached the capital. A determined propagandist onslaught, in which the Bolsheviks cooperated with the Mensheviks and Social-Revolutionaries, undermined his troops. The railwaymen tore up the tracks and mass insubordination wrecked the whole enterprise. Kornilov himself was arrested, but released later in the year.

The episode – it was no more – brought the Bolsheviks a tremendous access of strength. Within a week they had majorities in the soviets of Petrograd and Moscow. Trotsky, on his release from prison, was elected chairman of the former. This pattern was followed in many local and provincial soviets. In the countryside also, as peasant disorders spread, Bolshevik influence grew. The number of Party members increased some tenfold to 200,000 between January and August 1917.

Revolution was now on the agenda. The moment anticipated in Lenin's 'April Theses' was approaching rapidly. Lenin himself wrote on three successive days in mid September to the Party's central committee, urging that the moment had come. At the end of the month he moved from Helsingfors to Vyborg, nearer the Russian frontier. 'History will not forgive us if we do not seize power now.'

On 20 October (10 October old style) the supreme decision was taken. Lenin emerged from hiding to take part in the debate. By a majority of ten to two – Lenin, Trotsky, Stalin, and seven others against Zinoviev and Kamenev – it was resolved to initiate an armed insurrection. On 29 October (16 October old style) the Petrograd Soviet established a Military-Revolutionary Committee, under the presidency of Trotsky by virtue of his chairmanship of the Soviet. It included forty-eight Bolsheviks, fourteen left Social-Revolutionaries, and four Anarchists.

A week later the Committee had its plan ready. Insurgent troops, numbering some 20–25,000, would be used to occupy the key points in the capital. The date for the uprising was fixed for 6–7 November (24–25 October old style), the day before the second All-Russian Congress of Soviets was scheduled to meet.

Early in the morning of the crucial day the Central Committee, except for Lenin, Zinoviev, and Stalin, met for the last time before the night of decision. Their headquarters were at the Smolny Institute. It had been in former days a convent and school for young ladies; now it resounded to the final orders for the uprising. Trotsky assigned to each of his colleagues responsibility for supervising key services – food supplies, posts and telegraphs, railways, and liaison with Moscow. Trotsky himself directed the overall strategy of the *coup*. Lenin, back in hiding, once again urged immediate action. 'We must at all costs, this very evening, this very night, arrest the Government. . . .' Later that night, disguised by a bandage wound round his face, Lenin arrived at Smolny.

Early the next morning revolutionary troops and Red Guards went into action. They met almost no resistance. They methodically occupied one key-point after another – the railway stations, the power station, the telephone exchange, the State Bank, the bridges over the Neva. It was almost bloodless.

Karensky fled in a car placed at his disposal by the United States Embassy. He hoped to find troops to lead an attack against the revolution. This was about ten in the morning. At about the same time, Trotsky's posters appeared on the streets: 'To the citizens of Russia – the Provisional Government has been overthrown.' This was not quite true. The Winter Palace was still occupied by those ministers who remained, guarded by about a thousand officer cadets and a women's battalion. Not until early the next morning had enough Red Army soldiers infiltrated the enormous building and its grounds. Kerensky's ministers were finally arrested by Antonov-Ovseyenko, a man wearing pince-nez and a broad-brimmed hat. To the last the *coup* was bloodless. But it still remained a *coup*. Before the Bolsheviks held firm power, much blood would be shed.

CHAPTER 16

The Bolsheviks Conquer Power, 1917–20

So far the Bolshevik *coup* in Petrograd had been exemplary and bloodless. In Moscow, fighting lasted about a week. But in the capital it had not interrupted the opera or the ballet, the theatres, schools, or such government offices as were still functioning; and what met the eyes of the British Ambassador, Sir George Buchanan, as he walked towards the Winter Palace on the afternoon of 7 November? 'The aspect of the quay was more or less normal', he remarked, 'except for the groups of armed soldiers stationed near the bridges.' Since the beginning of the year so many upheavals had swept over the capital that each one made less and less impression. The next morning, 8 November, it still seemed as if nothing had happened. Anti-Bolshevik newspapers appeared on the streets, reporting the arrest of the Government and Kerensky's flight. But they anticipated no more than a temporary reversal. The same views were held in Cadet and official circles.

To all such hopes the lie was being given in Smolny, while naval guns were still using blanks to bombard the Winter Palace. Late that night the second All-Russian Congress of Soviets met to the sound of continuing shell-fire; and this sound mingled with Menshevik denunciations of the *coup d'état*. The Bolsheviks and their supporters, the left Social-Revolutionaries, had a considerable majority in the Congress – about 380 out of 650 deputies – and this was sufficient to confirm in office an exclusively Bolshevik Government. There were, it is true, certain Bolshevik leaders, of whom the most prominent was Kamenev, President of the Central Executive Committee of the Soviet, who urged that Mensheviks and S.R.s be invited to share office with the Bolsheviks. They argued that a one-party Government would be able to maintain power only 'by means of political terror'. But this was anathema to Lenin and Trotsky. Had they seized power in order to share it with their opponents? Those opponents whom Trotsky now consigned to the 'dustbin of history'?

In any case, scruples such as those of Kamenev belonged rather to the future than to the present. And the present belonged indisputably and unchallengeably to the Bolsheviks. When Lenin, in his capacity as President of the newly formed Council of People's Commissars (as the Bolshevik Government was named), rose to address the Congress, he was greeted with an indescribable ovation. There he stood, in John Reed's description –

a short, stocky figure, with a big head set down in his shoulders, bald and bulging. Little eyes, a snubbish nose, wide generous mouth and heavy chin. . . . Dressed in shabby clothes, his trousers much too long for him. Unimpressive, to be the idol of a mob, loved and revered as perhaps few leaders in history have been. . . . [1]

These were the halcyon days of the Revolution. But, as the Russian proverb has it: 'Don't praise the day till evening comes.'

Lenin, in his first public appearance since his flight to Finland in July, would now read out the first two historic decrees of the Council of People's Commissars – the decree on peace and the decree on land. The first invited 'all the belligerent peoples and their governments to open immediate negotiations for an honest democratic peace', that is, a peace without annexations or indemnities. The Soviets also abolished all secret diplomacy and undertook to publish all the wartime secret agreements from which they would, Lenin declared, refuse to derive any territorial gains. For the general purpose of the decree, Lenin proposed an immediate armistice to last not less than three months. He ended with an appeal to the working classes of Britain, France, and Germany to support the Soviets' peace policy. This was a disguised appeal for a European revolution, belief in which had been one of Lenin's foremost motives in urging on a Russian revolution in September and October.

1. John Reed, *Ten Days That Shook the World*, New York, 1960 ed., p. 170.

After peace came land. Lenin's second decree abolished without compensation and nationalized all private property in the form of landowners' estates, and appanages belonging to the Crown, monasteries, and the Church. Local land committees and peasant soviets would take into custody all such land, together with all livestock and implements, for distribution among their members. The decree prohibited the use of hired labour and also the sale, mortgaging, leasing, or alienation of the land. This was seemingly the most far-reaching agricultural reform in Russian history. In actual fact it merely recognized a *fait accompli*. The peasant seizure of the land had been in full spate for months before Lenin spoke and would continue throughout the early part of 1918. It was a movement, unchallengeable by any force that the Bolsheviks could conceivably have mustered. It expressed, as nothing else could have done, the dual nature of the October revolution. On one side, the Bolsheviks – on the other, the peasants.

In order to realize the Soviet State [wrote Trotsky] there was required a drawing together and mutual penetration of two factors belonging to completely different historic species: a peasant war – that is, a movement characteristic of the dawn of the bourgeois development – and a proletarian insurrection, the movement signifying its decline. That is the essence of 1917.

It is not necessary to accept Trotsky's historical framework to perceive the justice of this analysis. The Bolsheviks' attitude to the land question was probably the crucial factor in the Civil War. Could the White generals offer anything more than the return to a landlord economy? And who would fight for that? On the other hand, so many peasants returned from the towns to secure their share in the distribution that the average holding increased only very slightly. Land-hunger was by no means appeased, and Bolshevism found itself saddled with about twenty-five million smallholders working tiny plots by the most inefficient methods.

On the morrow of the revolution, Lenin's land policy was primarily of political importance in consummating the separation of the left Social-Revolutionaries from their mother-party.

The latter tended more to the right and to represent the in-terests of the larger peasant-proprietors. This accretion of Bol-shevik power and influence was of considerable help in the struggle with the Constituent Assembly. This threatened to be the main internal obstacle to the Bolsheviks. Like all Parties, the Bolsheviks had vigorously campaigned for the summoning of the Assembly. It had been one of their chief weapons in belabouring Kerensky that he was purposely delaying its con-vocation. Now, however, when the elections were held in mid November, the Assembly might show itself to be too much of a good thing. The Council of People's Commissars wanted no rivals for power.

The election results confirmed this fear. Less than half the electorate of more than ninety millions actually voted, so the results have to be interpreted with some caution. But it is still clear that the Bolsheviks were in a minority. They received about a quarter of the forty million votes cast and 175 seats. The left Social-Revolutionaries got forty, the right Social-Revolutionaries 370. There were fifteen Mensheviks, seventeen Cadets, and about eighty representatives of national groupings, most of whom were anti-Bolshevik. The pro-Bolshevik votes came from the industrial centres and the army, particularly those units stationed in Petrograd and Moscow.

Hardly had the Assembly met in January 1918 than it was dispersed by trigger-happy Red Guards, on the instructions of the Council of People's Commissar's. At the time, not many dogs barked. Lenin argued that the soviet form of democracy represented a higher type than mechanical and formal major-ities. Be that as it may, the forcible dissolution of the Con-stituent Assembly was not only a break with the great bulk of the revolutionary aspirations of the nineteenth century; it also widened the gap between the Bolsheviks and all the other left-wing Parties and groupings, apart of course from the left Social-Revolutionaries.

The Bolsheviks had now disposed of what constituted *as yet* their most serious internal danger. They now turned to the greatest external menace – the German and Austrian armies on the soil of what had once been the Empire of All the

Russias. At this time the German lines ran southwards from east of Riga to east of Lvov. They then sloped gently south-eastwards to the Black Sea, to include most of Romania.

Lenin's original peace decree had brought no response from either the Allied or the Central Powers. This did not prevent him from taking the policy a stage further. About a fortnight afterwards, the Bolshevik Government ordered fraternization on all fronts and instructed General Dukhonin, the Russian Commander-in-Chief, to propose to the Germans an imme-diate cease-fire. On the same day, Trotsky, the Soviet Foreign Commissar, informed the Allied Ambassadors in Petrograd of the Russian peace move. But Dukhonin rejected the order and the Allied Governments refused to entertain Trotsky's plan.

The Allied military missions accredited to Dukhonin's head-quarters at Mogilev (except for the U.S. mission) threatened that 'the gravest consequences' would follow any unilateral Russian violation of the treaty of 5 September 1914. Trotsky denounced this as intervention in Russian domestic affairs. 'The soldiers, workers, and peasants of Russia', he said, 'did not overthrow the Governments of the Tsar and Kerensky merely to become cannon fodder for the Allied imperialists.' Dukhonin was replaced by the Bolshevik, Krylenko. Krylenko left for Mogilev on 23 November. He took his time on the journey in order to depose *en route* as many hostile generals as possible. He did not arrive until 3 December. A few hours later Duk-honin was killed by his troops. The Allied military missions had left for Kiev, where White resistance to the Bolsheviks was already building up. Relations between the Allies and the Soviets were further inflamed through the Soviet repudiation of all Tsarist debts to foreign Powers;[1] and also through Trotsky's calculated publication of all the secret Allied agreements that he found in the archives of the Foreign Ministry. Matters such as the Anglo-Russian agreement of 1907 and the Sykes-Picot agreement of 1915 for the carve-up of Turkey's Arabian Empire seriously compromised and discredited the Allied

1. This undertaking had first been given in 1905 when the French made what was up till then their largest loan to Tsarist Russia.

cause, especially in the United States. Wilson's Fourteen Points could hardly redress the balance.

In the meantime, the Germans had agreed to negotiate with the Bolsheviks and a Russo-German armistice was signed on 15 December. The ensuing Treaty of Brest-Litovsk was finally ratified by the Fourth Congress of Soviets on 16 March 1918. This long interval was due in part to Trotsky's stalling tactics. He hoped to use the negotiations to stimulate revolution in Germany and Austria so as to take the weight off Russia. But nothing came of his efforts, apart from a wave of fairly widespread strikes. At one time there was also a Russian hope that aid might be forthcoming from the Western Powers. But this was even more of an illusion. In fact, Trotsky already thought that the Allies and the Germans might be collaborating against the Russians, though this did not mature until a year or so later. In the end, the Soviets had no choice but to accept a truly draconic peace. They had to yield up Estonia, Latvia, Lithuania, and Russian Poland to Germany and Austria, to recognize the independence of the Ukraine, Georgia, and Finland, and to evacuate the areas of Kars, Ardahan, and Batum in favour of Turkey. In addition, reparation payments of 6,000 million marks had to be made. In concrete terms, this meant that Russia lost one-third of its agricultural land and its population; more than four-fifths of its coal-mines; over half its industrial undertakings. Geographically speaking, Russia was pushed back from the Black Sea, and virtually cut off from the Baltic. It was almost the Grand Duchy of Muscovy over again.

No wonder the acceptance of these terms gave rise to the most violent polemics inside the Bolshevik Party and the Government! In the crucial Central Committee meeting of 23 February, the voting went seven for acceptance, four against, and four abstainers. A left Communist group led by Bukharin and Radek split off from the main Bolshevik Party so as to be free to urge forward a campaign for a revolutionary war against the Germans. Similarly, the left S.R. members of the Council of People's Commissars resigned their posts and denounced the treaty as a 'betrayal of the international proletariat and of the socialist revolution begun in Russia'.

To all such arguments Lenin reiterated his complete and utter disbelief in any further Russian capacity to fight. Moreover, to attempt to fight would, he believed, jeopardize the revolution. 'Germany', he said, 'was only pregnant with revolution.' But to the Bolsheviks quite a healthy child had been born – 'which we may kill if we begin war'. Who would risk a live child for one as yet unborn, all the more so as Russia could act as the German *accoucheur*? This was Lenin's conclusive argument. Russia sacrificed space in order to gain time. Russia won a breathing-space that would enable it to consolidate the revolution in preparation for the imminent struggle for the world. Russia accepted Brest-Litovsk in the same spirit in which the Germans had once accepted the Peace of Tilsit – 'and just as the Germans freed themselves from Napoleon', wrote Lenin in *Pravda*, 'so will we get our freedom.'

A curious situation now arose. In order to help stimulate revolution in Germany, Adolf Joffe, the first Bolshevik ambassador in Berlin, was actively in touch with the German left. He aided them with pamphlets and funds. In Moscow (whither the Soviet capital had been transferred in March) these activities were held to be so important that the reparation instalments due to Germany continued to be paid even *after* the German request to the Allies for an armistice. The aim was to minimize any danger of a Russo-German rupture that might unseat Joffe.

Similarly, the German Ambassador to Moscow, Count Mirbach was exhorted to continue to give financial support to the Bolsheviks.

Please use larger sums [he was adjured by his Foreign Minister in May 1918] as it is greatly to our interests that Bolsheviks should survive. . . . As a party, Cadets are anti-German; Monarchists would also work for revision of the Brest peace treaty. We have no interest in supporting Monarchists' ideas, which would reunite Russia. On the contrary, we must try and prevent Russian consolidation as far as possible and, from this point of view, we must therefore support the Parties furthest to the left.[1]

This analysis is as understandable as it is fallacious. The Bolsheviks were the very antithesis of disorder and chaos. The

1. Op. cit., ed. Zeman, Document No. 129.

Bolsheviks were rapidly developing the two characteristics of a stable State – a police force and an armed body of men, in other words, the Cheka and the Red Army. The one defended the régime against its internal, the other against its external, enemies. Censorship of the Press was introduced in March 1918. Also, of course, the Government was rapidly taking over control of the country's economic life. All large-scale industry was nationalized in June 1918. The model here was what Lenin called 'the state capitalism' of wartime Germany. Labour discipline and duties were enforced, strikes being declared treason to the State and the industrial equivalent of mutiny in the army. Private trade and banks were nationalized. Most important of all, in the spring of 1918 the State formed 'Food Armies' in order to requisition grain for the urban workers and Red Army.

All these developments stemmed originally from the need to consolidate the revolution. But they were given tremendous impetus by the forces of counter-revolution and of foreign intervention.

During 1918 the two movements became inextricably intertwined. At first intervention aimed at helping those forces opposed to the Germans. But as the White generals were at the same time opponents of the Bolsheviks, the two movements inevitably coalesced into one. Even after the armistice of 1918, when there could no longer be any question of reviving an Eastern front against the Germans, intervention was not simply anti-Bolshevik *tout court*. Each of the interventionist Powers – Great Britain, France, the United States, and Japan – had its own axe to grind, and this disunity helped the Bolsheviks to survive. During the whole campaigning period of some two years, the Bolsheviks held fast to the historic heartland of Russia, the territory around Moscow and Petrograd. They were thus able to concentrate their forces at any threatened point on the periphery. On the other hand, the Whites had the advantage of surprise and also of access to the sea.

The first White Army formed in the territory of the Don Cossacks. It was led by Generals Kornilov, Denikin, and Alexeyev, the former Tsarist Chief of Staff. But it was not from here that the first blow came – it was from the Czechoslovak Legion of about 30,000 men.

This was formed originally of prisoners taken from the Austro-Hungarian Army. The idea was that they should be evacuated from Russia to join the Allied forces fighting against the Germans. But in view of the Bolshevik assumption of power and the consequent collapse of the Eastern front, the Entente, on which the Legion was financially dependent, hoped to be able to use the Legion inside Russia – whether against the Bolsheviks or the Germans is not clear. In this state of indecision and confusion, with the Legion strung out along the Trans-Siberian Railway, a rumour that it would be disarmed and shipped back to Germany was enough to precipitate a full-scale revolt. Within a matter of weeks, a row of towns from Samara on the Volga to Vladivostok on the Pacific coast was in Czech hands. Had the Legion been able to cross the Volga, they might have swarmed across the unprotected plains to Moscow. In actual fact they were stopped at Kazan, where Trotsky, now Commissar for War, personally intervened in the battle, rallying the demoralized and retreating Red forces.

This was a strictly limited defensive victory. It did not prevent American, Japanese, and small British, French, and Italian detachments landing at Vladivostok. The pretext was: 'aid to the Czechs'. Thence the interventionists moved westwards into Siberia, where, with British and French support, a Government under Admiral Kolchak, the former commander of the Black Sea Fleet, was established.

In the meantime, Allied troops were landing at the White Sea ports of Murmansk and Archangel. Simultaneously, an anti-Bolshevik Government consisting mostly of Social-Revolutionaries was set up in Archangel under the leadership of Chaikovsky, the former Populist. A White Russian Government was set up under British aegis at Ashkabad in Turkestan. The British troops had advanced through Baluchistan and Persia, and occupied the Transcaspian area. By August 1918 about twenty different Governments were functioning on Russian soil.

Superimposed on the turmoil, that same summer, was an anti-

Bolshevik revolt of the left S.R.s. They hoped to embroil the country in a renewal of the war against Germany. A series of assassinations gave the signal: that of Count Mirbach, the German ambassador; of Uritsky, the head of the Petrograd Cheka; and an attempt on Lenin's life by a Jewish girl, Dora Kaplan. Associated risings took place in Moscow and Yaroslavl. The conspirators were helped by the French Ambassador. But the attempt failed. Its sole sequel was to intensify the terror wielded against the enemies of the régime.

The main threat to the Bolsheviks still came from outside the country. It was in these circumstances that Chicherin, a former Menshevik, who followed Trotsky as Commissar for Foreign Affairs, approached the Germans with a proposal for their collaboration in parrying the Allied forces in the north. The Germans, Chicherin proposed, should advance to prevent a possible southward advance of the British in Murmansk, while the Soviets withdrew their troops around Vologda in order to protect Moscow. But nothing came of this. The Germans by now had troubles enough of their own.

The end of the war in the West to some extent freed the hands of the Entente in its dealings with the Soviets. In December 1917 an inter-Allied agreement at Paris, signed by Milner and Lord Robert Cecil for England and by Clemenceau and Foch for France, had divided Russia up into Franco-British spheres of interest. The French were allotted Bessarabia, the Ukraine, and the Crimea. To the British fell the Cossack lands, the Caucasus, Armenia, Georgia, and Kurdistan. Pursuant to this agreement, both Powers sent ships to Novorossisk with arms for Denikin. Later, a French naval division landed at Odessa and British troops moved into Batum and Baku, attracted, it has been said, 'by the smell of oil'. Other White generals to be helped by the Allies were Miller in north Russia and Yudenich in the west.

As the war in the West ended, with the Soviets hemmed in on all fronts and under increasing Allied pressure, Lenin expected a concerted attack by world capital. The answer to world capital was world revolution. This was not an unreasonable gamble on empirical grounds alone, to say nothing of its theoretical justification in Marxist eyes. Habsburg and Hohenzollern fell. Soldiers' and Workers' Councils spontaneously formed themselves in Germany on the Russian model. *Pravda* proclaimed on 1 November: 'The world revolution has begun.... Nothing can hold up the iron tread of revolution.' A month later, Radek, the foremost Soviet journalist, predicted that the spring of 1919 would find all Europe communist. Zinoviev, the President of the Communist International, looked forward to transferring the headquarters of his organization from Moscow to Paris.

The Soviets combined revolutionary propaganda with offers of surrender. In a note of 4 February 1919, Chicherin offered to meet Russia's foreign debts, to lease concessions, and to give up territories to be occupied by armies drawing their support from the Entente. Russia, should these terms be accepted, would then undertake not to intervene in the internal affairs of the Powers. But they were not accepted. The Bolsheviks were also unable to take advantage of President Wilson's attempt to mediate between themselves and the White forces at a meeting to be arranged on the Island of Prinkipo. The French encouraged the Whites to turn down Wilson's invitation.

In these circumstances, revolutionary propaganda continued unabated. Germany was the focal point of Russian hopes. It was here that Lenin made his greatest effort. Ill-nourished as Russia was, he still set aside a store of grain for transport to a Germany suffering under the Allied blockade. He also assembled a small delegation of diplomats to represent Russia at the Berlin Congress of Soldiers' and Workers' Councils. But the new German Government, predominantly Social-Democratic, refused to admit both the grain and the diplomats. The Russians remained as isolated as ever. Not even the formation of the third Communist International in March 1919 helped much. Its fiery proclamations and its optimistic prognoses could not obscure the fact that it was quite unrepresentative of the Western-European labour movement. The Soviet Republics in Hungary and Bavaria were merely passing phases.

In 1919, as in 1918, the Soviets had to win through on their own. In one way they were in a weaker position because they were now blockaded by the Allies; also, the Allies, by Point 12 of the Armistice agreement with Germany, tried to ensure that the German troops on the Eastern front would return only to 'within the frontiers of Germany as soon as the Allies shall think the moment suitable, having regard to the internal situation of these territories'. (This condition was repeated *vis-à-vis* the Baltic countries in Article 433 of the Versailles Treaty.)

In early 1919 Kolchak, to whom the Allies looked to form the future non-Bolshevik Government of Russia, moved westwards across the Urals in the direction of Moscow. This was the opening move in a year of incessant campaigns that would reach its climax in the autumn. At first Kolchak's three armies rapidly advanced. Ufa fell and then, in April, Samara and Kazan were menaced. The bulk of the Red Army was in the south, holding the line against Denikin. But by the end of April Kolchak had reached his farthest point westwards. His southern wing, with its over-extended external lines, was outflanked in a wide sweep ordered by S. Kamenev, the Soviet commander on the eastern front and one of the 30,000 former Tsarist officers mobilized by Trotsky for service with the Red Army. Kolchak was forced back from Ufa and eastwards across the Urals. He lost pitched battles at Chelyabinsk and Omsk. By the end of the year, Kolchak's troops had simply disintegrated and he himself had fallen into Red hands. He was executed early in 1920.

Hardly had the main threat from Kolchak been repelled than the southern front under Denikin came alive. Here a virtual state of chaos prevailed. Guerillas, partisans, peasant anarchists under Makhno, and Ukranian nationalists led by Petlura, were all indescribably and inextricably intertwined. The peasantry had no love for the Bolsheviks, and Denikin at first made good headway. By the end of June he had taken Kharkov and Tsaritsyn (now Volgograd), and Kiev by the end of August. By mid October Denikin was in Orel, a bare 250 miles south of Moscow. At the same time, a diversionary attack was launched by General Yudenich against Petrograd. His base was in Estonia and he enjoyed British advice. This was the second such attack. The first had been thrown back in May under Stalin's leadership. Soon Yudenich was in the suburbs of the city and, but for the railway link with Moscow, had cut it off from the outside world. So critical was the situation that Lenin contemplated abandoning Petrograd and withdrawing to Moscow. This time Trotsky rallied the defenders. Yudenich was not strong enough to besiege the city, and since he could also not advance, he had no choice but to retreat. A month later he was back in Estonia, where his army broke up.

This reverse severely weakened Denikin in the south. He was no more popular with the local population than the soviets. Like that of all the White armies, his administration was marked by corruption, brigandage, and political reaction. The end of the year saw him inexorably on the retreat. He withdrew, with British aid, to the Crimea, and turned over his command to General Wrangel.

But Wrangel had no hope of resuming the struggle. All the more so as both the British and French had pulled out of the Black Sea and Transcaspian area and were preparing to lift the blockade. This was in fact withdrawn in January 1920.

The last throw in the civil war came from Poland. In April 1920 Pilsudski advanced eastwards into the Ukraine, and on 7 May took Kiev. Here he received a congratulatory telegram from King George V. But a Ukrainian revolt, on which Pilsudski had counted, did not materialize. The telegram was also of no great help. A swift turn came in the tide of battle. The Poles had to evacuate Kiev, and soon the Russians, carried forward on a swell-tide of traditional patriotism, were on the River Bug, the rough ethnographical barrier between Russian Ukraine and Poland proper. It was at this time that Brussilov, the former Tsarist Commander-in-Chief, offered his services to the Red Army. Lenin, desperate at the continued isolation of the Russian revolution, now gambled for the highest stakes. He would take the revolution Westward by force of arms, 'probe Western Europe with Russian bayonets', as he put it. The target was Germany.

At this time the Second Congress of the Comintern was in session and the delegates could see pin-pointed on a map the advance of the Russian armies.

With a clutch at their hearts, [said Zinoviev] the best representatives of the international proletariat followed the advance of our armies. . . . We all understood that on every step forward of our Red Army there literally depended the fate of the international proletarian revolution.

In the event, Lenin's gamble did not come off. The Russians were thrown back before Warsaw and were soon themselves on the retreat. Early in 1921 the Treaty of Riga between Russia and Poland brought an end to the war in the West. The Bolsheviks had maintained themselves in power; and the non-communist world had perforce to acknowledge failure. The process took longer in Transcaucasia, Central Asia, and the Far East. But by February 1921 three anti-communist republics in Georgia, Armenia, and Azerbaijan had fallen. In the same year Khiva, Bokhara, and Turkestan fell to the Bolsheviks, and at the end of 1922 Red troops at last entered Vladivostok, the last stronghold of the Whites.

CHAPTER 17

Economic Experiments

WHEN the Bolsheviks seized power in October 1917, they found a country on the verge of economic collapse.

The cultivated area had been reduced by the war by a sixth, the number of horses available for agriculture by nearly a third, the cereal harvest was down by fourteen per cent. . . . The product of industry was a little more than three-fourths of what it had been in 1913. The railway system had suffered severely from the strain of war, and from the lack of replacements and repairs. . . . The money in circulation was twelve times as much as in July 1914. In the country, the paper rouble was worth from a tenth to an eighth of the pre-war rouble. . . . A fall in real wages, calculated at twelve to fifteen per cent, had occurred: and the cost of living in October 1917 was five times that of a year earlier.[1]

This was no favourable background for the construction of socialism. But far worse was to come.

The Bolsheviks had at first no intention of aggravating these conditions by any drastic intervention in the Russian economy. The aim was simply to keep what industry there was in operation and to control such key sectors of the economy as would prevent any reversion to the previous régime. The nationalization of the banks, of key enterprises engaged on war work, and the State monopoly of the grain trade did not go much beyond the economic policies of the belligerent Powers in the First World War. Not until May 1918 did the Bolsheviks nationalize a whole industry. The first was sugar, then came oil, and then the establishment of State monopolies in foreign trade and in such commodities as spices, yarn, coffee, and matches.

This change in policy stemmed from two main causes. First, an effort had to be made to curb anarcho-syndicalist tenden-

1. Sir John Maynard, *The Russian Peasant and Other Studies*, London, 1942, p. 83.

cies. The decree on Workers' Control of November 1917 gave factory committees of employees the right to supervise the management and to enjoy access to correspondence, records, etc. But this was only a limited right by comparison with the owners' retention of full control over the conduct and administration of the enterprise. It was at this that many factory committees jibbed, frequently taking factory administration into their own hands. In this way workers' administration at length forced the State to enter on a policy of widespread nationalization.

The second cause was more general. Intervention and civil war accelerated and intensified developments in the economic as much as in the political field. If it brought with it the further development of terror as a political weapon and the suppression of non-Bolshevik groups and opinions, it also entailed a tightening of the State's control over the economy. The Bolsheviks were at war, and centralized control in the interests of war production was an obvious recourse – again on the pattern of the economies of the belligerent Powers. From the summer of 1918 onwards, nationalization went rapidly ahead. It began at the end of June with a decree of General Nationalization which prepared the way, from the legal point of view, for the nationalization of almost all large-scale enterprises, no matter in what industry or service.

This was the preliminary to what is known as 'war communism'. The exigencies of civil war forced on the Bolsheviks measures of State control over every branch of the national life. They conscripted all private and public wealth, as well as manpower; they banned all private trade; they assumed powers to direct labour wherever it might be most needed – in factories or the armed services. Wages were paid in kind, since paper money had lost virtually all value; and they acquired compulsory powers to requisition foodstuffs and grain from the peasantry.

This was essentially an *ad hoc* emergency policy; undertaken purely and simply in order to survive. Precisely how *ad hoc* it was may be judged from Kamenev's admission in December 1921: 'The proletarian State was unable to adminis-

ter and shape as a Socialist organization the property which
it was compelled to seize.' In Lenin's words, war communism
was 'dictated not by economic, but by military needs, considerations, and conditions'.

A further complication came from the position of the
peasantry. The peasant revolution of 1917, however welcome
to the Bolsheviks for its disintegrating effect on the old bourgeoisie and landowners, was no part of their calculations. It
was no part of Bolshevism to create and accept a system
whereby a class of twenty to twenty-five million smallholders
were tilling tiny plots with scarce and primitive equipment.
Here was a true anomaly. War communism emphasized the
contradiction. The urban workers' alliance with the peasantry
remained as urgent as ever – if not more so. Yet it had to be
strained to the uttermost, if not actually broken entirely,
through the need to force grain into the towns. This took place
at a time when the cultivated area had decreased, with a consequent decrease in the size of the agricultural product. When,
in 1919, a Government decree brought the millions of peasant
allotments under State control and carried out the policy
of requisitioning all produce surplus to the needs of subsistence, and the peasants found themselves with no incentive to produce anything more than they themselves could
consume.

The effects of this policy, combined of course with the devastation and destruction of seven years of war, soon showed
themselves in every aspect of agricultural life. By 1921 stocks
of cattle were less than two-thirds of their 1913 total; sheep
were about fifty-five per cent, pigs forty per cent, horses
seventy-one per cent. The sown area was just under half of
what it had been in 1913; the same held good for the total of
grain crops (which includes potatoes).

The industrial picture was even worse. By 1920, in the textile industry, only six per cent of the spindles were in operation, as compared with pre-war; the steel industry produced
less than five per cent of its pre-war output; the Donetz coal-
mines ten per cent. Manufactured consumer goods amounted
to about thirteen per cent of their pre-war total. An almost

complete breakdown in transport brought all semblance of normal life to a standstill.

Hand in hand with economic collapse went the depopulation of the larger towns. Moscow and Petrograd each lost several hundred thousand inhabitants. A pitiful mass flight to the countryside took place, where the food situation was relatively easier. Those workers who remained in the towns fell victim to every form of demoralization. Productivity fell catastrophically, as wages covered a bare fifth of the cost of living and the worker had to spend at least half his time trading on the black market. The goods he dealt in were those he stole from his place of work. Small wonder that Lenin told the Tenth Party Congress in March 1921: 'The poverty of the working class was never so vast and acute as in the period of its dictatorship. The enfeeblement of workers and peasants is close to the point of complete incapacitation for work.' The human tragedy reached its climax in an unprecedented drought in the lower Volga region in 1920–1. This caused the death of about five million people. Altogether, the seven years of war, civil war, famine, and starvation left Russia with a population deficit of about twenty-eight million.

The political consequences of this collapse had been steadily becoming more and more evident since the autumn of 1920. Peasant disturbances had begun in September with the first releases from army service. They gathered strength all through the autumn when, as Lenin admitted, 'tens and hundreds of thousands of disbanded soldiers' were taking up arms against soviet society by way of banditry, highway robbery, and political revolt. At the beginning of 1921 the Tambov area was for a few months completely in the hands of peasant guerilla forces, led by a former Social-Revolutionary.

The fact is that the alliance of workers and peasants which had, albeit for differing reasons in each case, carried the Bolsheviks to success in the civil war, was not only breaking up but was also leaving the Party isolated. The peasants had been quite willing to support the Party so long as it was a question of retaining the land. But they were otherwise politically apathetic, and in a state of near-revolt at the policy of grain requisitioning.

The existence of a cleavage between the industrial workers and the Party came to the fore in the Workers' Opposition. This claimed a greater role for the proletariat in the work and control of industry. It opposed Trotsky's policy of forming labour brigades of militarized workers and putting the railway trade unions under martial law.

Finally, there came, in Lenin's words, 'the flash which lit up reality better than anything else'. This was the Kronstadt revolt of March 1921.

It began early in the month with a demand for 'Soviets without Communists' in which peasant interests were prominent. Many of the Kronstadt sailors and Red Army men who led the revolt were in fact recent drafts from the countryside and brought with them the current political attitudes. They demanded, in essence, more freedom for the peasant and for the small-scale industrial producer. Politically, the movement aimed at breaking the Bolshevik monopoly of power, the Bolshevik dictatorship, by securing freedom of speech, Press, and assembly for the trade unions, left-wing socialists, and peasant organizations. It was supported not only by Anarchists and Mensheviks, but also by Bolsheviks. There was a call for a 'Third Revolution'.

The revolt broke out a few days before the opening of the Tenth Party Congress. Trotsky ordered the insurgents to surrender. They refused. At once stern military measures were taken. Loyal troops, led by Tukhachevsky and spurred on by Trotsky, advanced, camouflaged in white overalls, across the frozen Bay of Finland. Had Trotsky not acted quickly, the ice might have melted and made a land-based attack impossible. As it was, the fighting, in bitter, blinding snow storms, lasted for ten days. A measure of the crisis is the fact that many of the Party delegates in session were sent from Moscow to Kronstadt to participate in the final storming of the base.

The immediate crisis passed, but it left with Lenin the conviction that the alliance with the peasantry – the alliance that had been the very foundation of the revolution – must be reestablished. Late in February 1921, Lenin had already submitted to the Party Central Committee a project for a new economic policy. A few weeks later this emerged, after due discussion by the delegates, as the starting-point of an economic programme that entirely scrapped the policies of war communism. It began primarily as an agricultural measure that would give an incentive to the peasant to produce more food for the towns; it then broadened out into a medium for the development of commodity exchange between town and country; and, lastly, into an encouragement to industrial productivity.

At the back of it all was the necessity of saving the revolution. This was the point that had been reached. It had always been held by Lenin that in Russian conditions – that is, in a country where bourgeois capitalism had never developed – there could be no direct transition to socialism. The accomplishment of a socialist revolution would depend, Lenin argued, either on a socialist resolution in the more advanced countries of the West or on 'a compromise between the proletariat which puts its dictatorship into practice or holds the State power in its hands, and the majority of the peasant population'.

In the absence of the first of these conditions, the second automatically moved into first place. To this necessary compromise Lenin's New Economic Policy (or N.E.P. as it soon came to be known) made its own specific contribution.

What it did initially was to replace the requisitioning of the peasants' surpluses by a graduated, agricultural tax in kind (from 1923 onwards in money only). This latter was calculated as a proportion of the surplus left over after providing for the minimum subsistence needs of the peasants' family and dependants. The new policy did indeed contain a residue of the previous requisitioning element. But it had a vital difference in that, since it took only a fixed proportion of the surplus, it gave every incentive to the peasant to produce to the maximum and thus to increase *his* share of the surplus; or, as the official decree put it – 'every peasant must now realize and remember that the more land he plants, the greater will be the surplus of grain which will remain in his complete possession.' In 1925 the wealthier peasant, the *kulak*, was further favoured in being allowed to use hired agricultural labour.

But what would happen to the peasants' surplus? It was here

that N.E.P. developed its second arm. It implied that there was a right of free trade in agricultural produce. In other words, the market was re-created. It was at first limited to local fairs and marts. This limitation later fell away and a class of so-called *Nepmen* came into existence. They dominated retail distribution and also played an important role as wholesalers. In 1922–3, for example, private traders controlled about seventy-five per cent of the retail trade of the Soviet Union. The share of State and cooperative trading respectively was about fourteen per cent and ten per cent. The *Nepmen* functioned as the medium through which trade between town and country was re-established.

The New Economic Policy signified in essence a Communist going to the peasant Canossa, a partial surrender to the vast peasant majority. On the other hand, the State retained what Lenin called 'the commanding heights' of the economy. These were primarily the largest industrial installations. According to a census of 1923, although quantitatively the share controlled by the State was comparatively small – of 165,781 enterprises eighty-eight per cent belonged to private individuals, 8·5 per cent to the State and three per cent to cooperatives – *qualitatively* the share of the State enormously outweighed that of its partners. Thus, the eighty-eight per cent privately owned enterprises employed only 12·5 per cent of the total number of workers employed in industry, whereas the State-owned enterprises employed just over eighty-four per cent of all industrial workers. Put in another way, the average number of workers in a State-owned enterprise was 155; in cooperative and private enterprises it was fifteen and two respectively. The latter, it is clear, was not much more than a tiny workshop or something similar. Altogether, private enterprises accounted for only five per cent of gross production.

Moreover, the State controlled all such ancillary services as credit and banking, transport, and, of course, foreign trade. All these remained State monopolies. The overall consequence was a strangely mixed economy in which the State virtually monopolized industrial life, whereas agricultural production was controlled by some twenty-five million small producers. This was an inherently unstable mixture, all the more so as the peasant producer held the whip hand over the industrial proletariat. Moreover, the peasant had only the obligation of a taxpayer to the State. The town worker, on the other hand, was exposed to all the hazards of rationing and unemployment. At this time the quip was current that the initials N.E.P. denoted 'New Exploitation of the Proletariat'. By early 1924 there were nearly one and a quarter million unemployed. At its height the total was to reach two million.

Until 1928 there was no basic change in this mixture of competing economies. It was a system that Lenin termed 'State capitalism'. It would not have been difficult at this time to envisage a Russian economic future formed of a growing and thriving class of merchants, a growing class of wealthier peasant farmers, with the State controlling and planning industry. This was in itself a highly unstable compound, and also an uninviting prospect. No wonder, therefore, that it gave rise to the most acute tension within the ruling Communist Party.

CHAPTER 18

The New Economic Policy and the Rise of Stalin

DESPITE all the penury and hardship, all the cleft between the dream of abundance and the reality of hardship, it still seems true that in many respects the twenties were the halcyon days of the revolution. In the social field there was steadily increasing expenditure on social insurance, sanatoria, workers' housing, hospitals, convalescent centres, and general welfare. Given the circumstances of the time, this was to be sure aeons away from socialism. But it was a beginning. Educationally, progressive methods in schools and the treatment of orphans and delinquents made Soviet Russia a centre of experiment and research. Even in the sphere of intellect, private publishing houses still continued to operate; non-Marxist historians still enjoyed positions of prominence and freedom of publication, provided, of course, that their work did not contradict the policy of the new régime. There was as yet no Government-inspired historical schema. In the arts, Trotsky was foremost in pouring scorn on the idea of a proletarian culture, and it was still possible for the 'Serapion Brotherhood', poets such as Mayakovsky and Esenin, novelists and short-story writers such as Isaac Babel, Pilnyak, Vsevolod Ivanov, and Leonid Leonov to choose and develop their themes with a broad degree of freedom – so broad, in fact, that in the cases of Pilnyak and Essenin, for example, their anti-industrialist leitmotive might strike at the roots of Soviet aspirations. Although many writers and intellectuals did indeed take to exile as escape from the Soviet régime, there was no profound break in cultural continuity.

The break with the past, intellectually speaking, was yet to come; and this was associated with the career of Stalin, the domination that the Party came to exercise, and the industrial and agricultural policies that were decided on.

Formally, of course, the Soviet Union was a democracy with

the whole population (apart from the disenfranchised bourgeoisie and adherents of the Tsarist régime) represented in the All-Russian Congress of Soviets. This was the pinnacle of many thousands of soviets throughout the Union, elected indirectly on the basis of one deputy for each 25,000 urban voters and one deputy for each 125,000 rural inhabitants. But the All-Russian Congress met for only about one week in the year and consisted of almost 2,000 members. Its Central Executive Committee of about 300 members was scarcely less cumbersome. Thus power devolved automatically on to the ten-member Council of People's Commissars elected from the Central Executive and the nearest equivalent to a Western European cabinet of ministers.

But virtually all the Commissars were members of the Communist Party and it was here, and particularly in the Politburo, that ultimate power had its abode. It was with this as a base that Stalin rose to dominant power, both in the Party and the country.

In the background stood two clashing elements: on the one hand, the renewed scope given to private enterpise in the N.E.P.; on the other, the continued isolation of Russia as the one communist Power in a capitalist world. Both these developments were as unexpected as they were unwelcome. The ban on the formation of opposition groups inside the Party was yet a third unwelcome factor. This was imposed at the Tenth Party Congress in March 1921, at the same time as the New Economic Policy was introduced. Lenin and Trotsky saw the immediate justification for the ban in the need to avert danger to the dictatorship of the proletariat. In the shadow of the Kronstadt revolt and the disruptive tension created by the concessions to the peasants – to say nothing of the weakened position of the proletariat – it was natural for the Party to close its ranks. Thus, the Congress prescribed 'the immediate dissolution of all groups without exception forming themselves on this or that platform. ... Non-fulfilment of this decision of the Congress must entail unconditional and immediate exclusion from the Party.'

There had previously been considerable freedom of grouping and discussion inside the Party. There had been an opposition movement in September and October 1917; a few weeks later, dispute had broken out on relations with the other left-wing Parties; the decision whether to accept or to reject Brest-Litovsk had again given rise to intense argument. The years of civil war had naturally seen a closing of the ranks. Even so, there had been a military opposition directed against the use of bourgeois military specialists; and in 1920 the Workers' Opposition had emerged to combat bureaucracy and centralist tendencies. In all these cases there had been full freedom to propose alternative policies. Thus, to outlaw the formation of internal groupings marked a significant turning-point.

This was all the more the case as it followed, by a few months, the final extinction of the Menshevik and Social-Revolutionary political activity. Until the autumn of 1920 these two Parties had still been able to lead a harassed but semi-independent life. By the spring of 1921, not only did no political body exist outside the Bolshevik Party, but also, within that Party, the expression of divergent viewpoints was severely restrained. It was not by any means repressed altogether. On the contrary, with the growth of Stalin's opposition to Trotsky, later joined by Zinoviev and Kamenev, there was, in the middle and later twenties, as much passionate controversy as ever before. All the same, the decision of 1921 signified an irrevocable turning-point. Its consequences coloured the whole of the subsequent history of Soviet Russia.

The sequel to this development was all the more nefarious in that it coincided with the transformation of the Party from an élite group into a mass organization. Lenin said in August 1917: 'if 150,000 landlords can rule Russia, why can't 240,000 Bolsheviks do the same job?' But by 1919 the 240,000 had become 313,000, by 1920 431,000, and by 1921 585,000. This growth accorded with the new tasks assumed by the Party. It had to provide leading personnel in the army (as political commissars), in the trade unions, in industry, and in every department of state, to say nothing of the manifold requirements of the cooperatives and local soviets. No wonder Party members were running into hundreds of thousands. As a corollary, the Party also developed its own apparatus, the men who lived from the Party, the *apparatchiki*. The instrument of power was becoming institutionalized. The transition was natural and probably inevitable. It corresponded, sociologically speaking, to the evolution characteristic of every large organization – the more members, the less individual power, and the greater the centralization of power. As a by-product of all this, the Party developed powers of appointment, promotion, and dismissal.

It was against this background – of tension generated through the operation of the N.E.P. in an ostensibly socialist society, of tension between a communist Russia and a capitalist world, and of tension between the free-thinking Bolshevik past and an increasingly centralized Bolshevik present – that Stalin rose to power.

At the end of the civil war he was Commissar of Nationalities – a post for which his Georgian origin made him eminently suitable – he was a member of the Politburo, and also Commissar of the Workers' and Peasants' Inspectorate. This last, known as *Rabkrin*, had been established in 1919, as a supervisory body over the Soviet Civil Service. Its job was to eliminate bureaucratic abuses. It worked through teams of workers and peasants who were sent to inspect the work of government departments. Stalin also acted as liaison between the Politburo and the Orgburo, the body that directed the activity of Party members as the interests of the State demanded. In 1922 his power was further enhanced by appointment to the post of General Secretary of the Central Committee of the Party. (His staff already numbered more than 600.) This was a further creation designed to coordinate the Party's top-level activities. But the general effect was something very different. The centre of gravity began to move away from the Politburo to the Secretariat, from the policy-making body to the body which supervised the execution of policy. The Politburo had originated at a time when the Party was an élite and when there could therefore be no problem in communication up and down the hierarchy. But when the élite became the mass and when the mass became burdened with a vast and complex burden of government in the worst conceivable circumstances, there was a proliferation of administrative bodies. These were Stalin's element and here was his *forte*.

Stalin's accretion of power went largely unremarked until the last year or so of Lenin's life. Stalin was as powerful as ever when Lenin died in January 1924. The 'political testament' in which Lenin proposed that Stalin be removed from office, was virtually suppressed at the subsequent meeting of the Central Committee. This had the support of Zinoviev and Kamenev.

The struggle for the succession to Lenin had already broken out in 1923: it now came into the open. Stalin was at first allied with Zinoviev and Kamenev against Trotsky. This first phase of the conflict began in 1923 and came to an end in January 1925. Trotsky was then forced to resign his post as Commissar for War and his presidency of the Revolutionary-Military Council. A few months later he was appointed to a post in the Council of National Economy, with largely nominal duties in the field of foreign concessions, electro-technical development, and an industrial-technological commission.

In the middle of 1925 Zinoviev and Kamenev in their turn broke with Stalin, and early the next year formed a bloc with Trotsky. As a result, Zinoviev was removed from the Politburo, and later from his post on the Executive Committee of the Communist International. At the same time, Trotsky and Zinoviev were likewise removed from the Politburo. The climax came in November and December 1927, when the three dissident leaders and seventy-five of their followers were expelled from the Party. Early in 1929 Trotsky was sent into exile and deported from Odessa.

Hardly had this, the so-called 'left opposition', been dealt with, than a 'right opposition' emerged in the Politburo. This was led by Bukharin, the editor of *Pravda* and secretary of the Comintern, Tomsky, head of the trade unions, and Rykov, Chairman of the Council of People's Commissars. By 1930 all three had been removed to lesser posts. Stalin now dominated the Politburo, seconded by Molotov, Voroshilov, and Kalinin.

The plethora of debate and controversy in these years surpassed anything in the Party's previous history. Here it cannot be followed in detail. Suffice it to say that it embraced every aspect of foreign and home policy. Had Stalin bungled the German revolution of 1923 and then the chances of the Chinese communists in 1927? Had he banked too much on the Anglo-Russian Trade Union Committee? Was the pace of industrialization too fast? Or too slow? Should the richer peasant be favoured? Or should he be taxed and coerced out of existence? Should intra-Party factions continue to be banned? Had the Party degenerated into a bureaucracy? Were the trade unions tied too closely to the State machine? Lastly, could socialism be built in one country alone – Russia – as Stalin argued, or was Trotsky's thesis of 'Permanent Revolution' the more valid analysis?

This political debate was of course strongly seasoned and suffused with a pure struggle for power in which the ostensibly political issues were manipulated as debating points for their public appeal. But this does not destroy their validity.

During these years, 1923–9, the whole Party line was in the melting-pot. What ultimately emerged victorious was a more or less monolithic Party, wedded to the doctrine of socialism in one country, to a policy of agricultural collectivization, and to industrialization as a means to the achievement of socialist abundance. Russia stood on the eve of a new revolution, even more tumultuous than that of 1917.

CHAPTER 19

The Second Revolution

THE intra-Party struggle of the middle and later twenties produced one novel doctrine – the doctrine of socialism in one country. This was Stalin's creation and contribution to the great debate. It signified a more or less complete break with everything that Bolshevism had previously proclaimed. The new theory decisively broke with the hitherto axiomatic view that socialism in Russia could be achieved only as part of an international, or at least a European, revolution. But this departure is unimportant. The point is that Stalin's discovery did in fact give a new *raison d'être* to the revolutionaries. They had seized power and grimly fought for power in the expectation of world revolution. But when this failed to materialize, Bolshevism found itself in a *cul-de-sac*. Furthermore, the revolutionary ebb was accompanied by some considerable improvement in the Soviet relationship to the capitalist world. This situation, however welcome as a tribute to the consolidation of Soviet power, was also embarrassing. Where do we go from here? This question inevitably arose.

Stalin's answer, therefore, that it was possible to construct socialism in Russia, leaving the international revolution to take care of itself – for the time being anyway – met a profound psychological need. It was a much more tangible concept than its only competitor – Trotsky's theory of permanent revolution. It gave new impetus as well as new direction to the whole of the subsequent history of Soviet Russia. Moreover, the implications of the concept do not end there; it also revived, and very strongly, the whole aura of Russian universalist messianism. Backward Russia would yet lead the world! This was the truly inspiring vision that Stalin was ultimately to evoke.

Stalin first propounded his doctrine, buttressed with a few selected gobbets from Lenin, in the autumn of 1924. Then, a year later, the Fourteenth Party Congress resolved that 'in the sphere of economic development, the Congress holds that in our land, the land of the dictatorship of the proletariat, there is "every requisite for the building of a complete socialist society" [Lenin]. The Congress considers that the main task of our Party is to fight for the victory of socialist construction in the U.S.S.R.'

By now the teaching of socialism in one country had imposed itself as the key tenet of Bolshevik doctrine. It had rapidly replaced the more far-flung aims of the revolution, and, as it were, nationalized them, rooted them in Russian soil.

But what did it mean in practice? What specific developments flowed from the doctrine? The answer in one word is – industrialization. As Marxists, the Bolsheviks were in any case committed to the belief in man's power to control his environment, to cast aside the chains of economic bondage, and to realize the freedom of abundance. This infallibly entailed industrialization, the means to the production of man's worldly goods. Socialists in the West could say with some show of justification that capitalism had solved the problem of production; it was now for socialism to solve the problem of distribution. But in Russia this theory was meaningless. There the problem was one of production first and last. Given the extremely low standard of living, the primordial requisite of any advance to socialism could only be industrialization.

But it would be absurd to pretend that this was the sole or even the most important motive of Russian industrialization. It was not undertaken, at least not in the first instance, for consumptionist purposes, but much more for the purpose of turning Russia into a great Power with a strong industrial base. This was nowhere argued more cogently or eloquently than by Stalin in a famous speech to industrial executives in February 1931, at the height of the First Five-Year Plan.

It is sometimes asked whether it is not possible to slow down the tempo a bit, to put a check on the movement. No, comrades, it is not possible! The tempo must not be reduced. . . .
To slacken the pace would mean to lag behind; and those who lag behind are beaten. We do not want to be beaten. No, we don't want to . . . [Russia] was ceaselessly beaten for her backwardness.

She was beaten by the Mongol Khans, she was beaten by Turkish Beys, she was beaten by Polish-Lithuanian *Pans*, she was beaten by Anglo-French capitalists, she was beaten by Japanese barons, she was beaten by all – for her backwardness. For military backwardness, for cultural backwardness, for political backwardness, for industrial backwardness, for agricultural backwardness ... You remember the words of the pre-revolutionary poet: 'Thou art poor and thou art plentiful, thou art mighty and thou art helpless, Mother Russia! ...'

We are fifty or a hundred years behind the advanced countries. We must make good this lag in ten years. Either we do it or they crush us. ...

In the somewhat calmer atmosphere of the post-war period, Stalin said:

The Party knew that a war was looming, that the country could not be defended without heavy industry, that the development of heavy industry must be undertaken as soon as possible, that to be behind with this would be to lose.... Accordingly the Communist Party of our country ... began the work of industrializing the country by developing heavy industry.

The first intimation of comprehensive socialist planning goes back to 1920, when the State Commission for Electrification was established in order to draft a programme for the electrification of the Russian Soviet Federated Socialist Republic. Then, in 1921, the State Planning Commission – Gosplan for short – was set up to draft a unified economic plan for the whole country. It started with a staff of forty economists and certain technical personnel. Their number increased rapidly and so did the scope of the work, with the establishment of many regional planning offices throughout the Union. But this was still far from constituting a plan in the modern sense. Gosplan conceived its task somewhat on the model of capitalist economic planning and limited itself in the main to the forecasting of trends and the analysis of the socialist trade cycle. Also, of course, Gosplan could not but point to the inherent contradiction in attempting to plan an economy in which the agricultural sector was totally unamenable to planning.[1] All

1. See pp. 283 ff. for a fuller discussion of this point.

the same, the control figures that Gosplan produced from 1925 onwards as an extrapolation of current trends, did in fact serve as a preparation for socialist planning. By 1927–8 they already filled a thick volume of 500 pages. The leitmotive of Gosplan thinking was that once the pre-war rate of production had been achieved, only limited progress could be expected thereafter.

But in December 1927 the whole leisurely tenor of Gosplan's activity was revolutionized. It was instructed by the Fifteenth Congress of the Communist Party to produce a five-year plan for the overall development of the Soviet economy. This followed the truly epoch-making Party decision of December 1926 that demanded 'the transformation of our country from an agrarian into an industrial one, capable by its own means of producing the necessary equipment'.

The same Party Congress of 1927 that authorized the drafting of the plan for industrialization also decided 'to pursue the offensive against the *kulaks*'. What was the connexion? Simply that no plan of intensive industrialization on the scale projected could conceivably be undertaken without bringing agriculture within its scope.

It must not be forgotten [said Stalin in referring to reduced grain deliveries from country to town] that in addition to elements which lend themselves to planning, there are elements in our national economy which do not as yet lend themselves to planning; and that, apart from everything else, there are hostile classes which cannot be overcome simply by the planning of the State Planning Commission.

Now the chickens of 1917 were coming home to roost. Had the Mensheviks been right in arguing that in such an economically backward country as Russia a proletarian and socialist revolution could not be carried through? An inevitable question as soon as the régime began seriously to deal with the problem of large-scale industrialization. A vicious circle existed. Apart from German credits, no worthwhile foreign loans or investments were available. Thus the imports of capital equipment essential to industrialization could be covered only by increased agricultural exports. But the type of land division that had been sponsored – willy-nilly – by the Bolsheviks in

1917 made any considerable increase impossible. That was the position in a nutshell, at least as it presented itself to the Soviet Government. Moreover, not only was it impossible to exercise economic control over some twenty-five million peasant households, but, apart from a small group of wealthier farmers, the vast majority of peasants had insufficient land to employ modern agricultural machinery profitably. In existing circumstances all that such peasants could manage was to market a very small proportion of their produce. They lived barely above subsistence level.

There were two basic reasons why the Government could not back the *kulaks*, the wealthier peasants. First, since the *kulaks* were the only group able to market substantial surpluses of grain, they stood for high prices, whereas the interests of industrialization required low prices to benefit the town workers. Second, such a policy would have given a class hostile to Soviet power even more influence than it already enjoyed. This conjunction of ideological and economic needs made the elimination of the *kulaks* inevitable.

Also inevitable, given the inefficiency of an agricultural system split up into millions of self-sufficient smallholdings, was the process of collectivization. This would serve the dual purpose of creating larger units producing marketable surpluses, and also releasing manpower for the demands of town and factory. The overall picture, as it presented itself to Stalin, is clearly evident in his remarks to a conference of students of agrarian problems.

Can it be said [he asked] that our overall peasant farming is developing according to the principle of expanded reproduction?... Not only is there no annual expanded reproduction in our small peasant farming, taken in the mass, but, on the contrary, it is not always able to obtain even simple reproduction. Can we advance our socialized industry at an accelerated rate, having to rely on such an agricultural base?... Can the Soviet Government and the work of socialist construction be, for any length of time, based on two *different* foundations – on the foundation of the most large-scale and concentrated socialist industry, and on the foundation of the most fragmentary and backward, small-commodity, peasant farming? They cannot. Sooner or later the end must be a complete collapse of the whole national economy. What, then, is the solution? The solution lies in enlarging the agricultural units, in making agriculture capable of accumulation, of expanded reproduction, and in thus changing the agricultural base of our national economy. But how are the agricultural units to be enlarged? There are two ways of doing this. There is the *capitalist* way, which is to enlarge the agricultural units by introducing capitalism in agriculture – a way which leads to the impoverishment of the peasantry and to the development of capitalist enterprises in agriculture; we reject this way as incompatible with the Soviet economic system. There is a second way – the *socialist* way, which is to set up collective and State farms, the way which leads to the amalgamation of the small peasant farms into large collective farms, technically and scientifically equipped, and to the squeezing out of the capitalist elements from agriculture. We are in favour of this second way.

The whole process of Soviet industrialization and collectivization was analogous to the enclosure movement in England – the creation of a more productive agriculture and the driving of the agricultural population into the towns of the industrial revolution. An even closer analogy is with the pattern of nineteenth-century Russian economic development. Then, the iron and steel industry, the oil-fields, the manufacturing installations, had been largely financed by foreign capital, the interest on which came from Russia's favourable foreign trade balance – and this depended on the export surplus of grain. The grain-producers, the peasants, were squeezed by high indirect taxes so that they had to sell their produce even though they themselves were forced to live below subsistence level. (There was a higher *per capita* grain consumption in the importing countries than in Russia, the country where the grain was actually produced.)

The Stalinist policy of industrialization and collectivization derived from somewhat similar principles. Its rationale owed something to the theories of the noted Trotskyite economist, Preobrazhensky. In his *New Economics*, published in 1924, he had argued, roughly speaking, that industry in such an economically undeveloped country as Russia could not, in the first instance at any rate, produce enough surplus to finance its further development. This would therefore have to be secured by squeezing agriculture and depressing the living standards of the workers – which is what actually happened. But Preobrazhensky had proposed fiscal measures and the manipulation of prices as the means to this end. There was no conception that industrialization might take place at the breakneck tempo that Stalin initiated.

THE BATTLE FOR AGRICULTURE

By 1928 a crisis was visibly approaching. Abroad, the war scare of 1927 had revived an ineradicable fear of attack.[1] At home, the political crisis came to a head with the exile of Trotsky to Soviet Central Asia. Economically, in both agriculture and industry, pre-war production levels had been reached. In normal circumstances, therefore, a much lower rate of growth would be expected than had characterized the previous six or seven years. To this it may be added that now that Stalin had defeated the main Trotskyite opposition he had to justify his victory, to say nothing of reaffirming his supremacy.

By now also, the private sector of the economy in trade and small-scale industry had been severely reduced through the Government's discriminatory measures. The stage was set for a further advance on the lines laid down by the Party Congress of December 1927, i.e. the planning and creation of a socialist industry and measures 'to restrict the development of capitalism in the countryside and guide peasant farming towards socialism'. Both these objectives were inextricably intertwined, as explained above. Neither could be achieved without the other, at least in the planners' view.

It is quite possible, of course, to imagine that the agricultural problem might have been tackled differently. In the natural course of events the development of individualist farming would have produced growing surpluses and those peasants dispossessed would have migrated to the towns. But the political objection to fostering the growth of private property, with – as its probable corollary – the emergence of a competing political

1. See pp. 304–5.

party, seems to have made this approach unacceptable. At any rate, it was never seriously tried. What actually resulted was enforced collectivization. The measure of this tumultuous upheaval can be gauged from Stalin's later admission to Churchill that the tension was as great as in any of the wartime crises.

The first moves towards collectivization were comparatively mild. At the end of 1927 the Party Congress proclaimed collectivization by example, and also imposed limits on the leasing of land by *kulaks* and their hiring of labour. Everything suggested that Stalin, as well he might, was approaching the peasant question very gingerly.

The first clashes came at the beginning of 1928. These were produced by a short-fall in grain deliveries to the towns. There is no reason to suppose that political motives had any part in this – merely peasant dissatisfaction with the prices fixed by the State. The Government replied with emergency administrative measures. Search parties were sent to the countryside to confiscate hidden stocks of grain, and the recalcitrant *kulaks* were imprisoned. At the same time, Committees of Poor Peasants were formed to denounce the hoarders. This encouragement of class war led to further violence in the villages and to a further reduction in the sown areas.

Even so, at the end of that year the first Five-Year Plan did not propose any radical change. In fact, its original version spoke frankly of 'the unusual difficulties [that] are involved in the problem of reorganizing farming on a collective basis. ... The fact must be frankly faced that in this field we are still feeling our way. ...' The object was to encourage the poorer peasants to enter collective farms, which would be favoured with financial and technical aid in the form of tractors and modern agricultural machinery; similarly, the middle peasant would be encouraged to improve his agriculture; and the *kulaks* would be crushed by additional taxes and other measures of financial discrimination.

This policy had at first some success. The number of collective farms between 1 June 1928 and 1 June 1929 rose from 33,000 to 57,000, embracing respectively 417,000 and more than a million peasant homesteads. But this success applied overwhelmingly to the poor peasants, those without land, horse, or cow. Those peasants who had something to lose – the *kulaks* and the less poor peasants – were unresponsive or hostile to the Government's plans. The autumn sowing of 1928 and the spring sowing of 1929 gave further evidence of the peasant intention to put pressure on the Government by restricting food supplies. This development was all the more unwelcome, intolerable even, as it coincided with an unexpectedly rapid increase in industrialization and the number employed in heavy industry.

Would industrialization be held up by an unregenerate, hostile peasantry? This was the prospect that opened up before Stalin. The answer came in the summer of 1929:

... We must *break* down the resistance of this class [the *kulaks*] in open battle and deprive it of the productive sources of its existence.... This is the turn towards the policy of eliminating the *kulaks* as a class ... the present policy of our Party in the rural districts is not a continuation of the old policy, but a turn ... to the new policy of eliminating the *kulaks* as a class.

Spurred on thus by Stalin, terror and repression came to the countryside. It was indeed 'open battle', akin to civil war. On the one side stood the power of the State, embodied in the dispatch of picked Party members to the countryside, and with the occasional use of Red Army units and police detachments. Also, the Government encouraged poor peasants and village soviets to seize from the *kulaks* machinery, cattle, and farm appliances for the benefit of the new collective farms. All this gained added bitterness from the urge to work off old grudges and resentments. The *kulaks* for their part retaliated by killing their cattle, burning their crops, and destroying their homesteads. All this they would rather do than let their property fall into the hands of the State. The frantic pace, but not the human tragedy, can be seen in the statistics: between 20 January and 1 March 1930, the number of collective farms almost doubled and the percentage of collectivized peasant homesteads rose from 21·6 per cent to fifty-five per cent. This went far beyond the totals envisaged in the Plan. In March 1930 Stalin himself called a halt to the turmoil. From now on, he declared, making an unprecedented *volte-face*, collectivization must be carried out on a voluntary basis and excessive socialization of property halted. The next few years did in fact see significant relaxations. When the collective farms had delivered their fixed quotas, they were allowed to sell surplus wheat, meat, vegetables, and fruit on the open market; and those peasants who so desired were allowed to withdraw from the collectives with their land and stock. In two months, March and April 1930, about nine million households took advantage of this freedom.

This was only a temporary retreat on the Government's part. By the end of 1932 there were again some fourteen million collectivized peasant households, more than half the total number. This followed in part from renewed governmental coercion, and in part from a certain relaxation in the collective system. The collectivized households, for example, were allowed to retain for family use their homes, small plots, cattle and poultry, and some small agricultural implements.

Hand in hand with the collectivization of agriculture went the establishment of machine-tractor stations. These had more or less a monopoly of agricultural machinery, with which they ploughed the collectives' land in return for a proportion of the crop. But their total was quite inadequate.

It was many years before Russian agriculture made its recovery from the turmoil and destruction of 1929–32. Not until 1934 did Stalin reveal the cost of the 'advance' to large-scale farming. There were thirty-three million horses in 1928 – and fifteen million in 1933. The respective figures for horned cattle were seventy million and thirty-four million; for pigs, twenty-six million and nine million; for sheep and goats, 146 million and forty-two million. This is to say nothing of the millions of *kulaks* and their families deported to forced-labour camps and new industrial locations beyond the Urals. In some areas the actual loss of life reached unimaginable proportions. This applied to the Ukraine, for example, where famine conditions prevailed in 1932 as a direct result of the disorganization of peasant agriculture. In Kazakhstan, where the depletion of livestock was probably greater than anywhere else (seventy-three per cent of the cattle, eighty-seven per cent of the sheep and goats, eighty-eight per cent of the horses) 'the number of Kazakhs ... was less by one million or more than the number that would normally have been expected in 1939'.[1]

THE BATTLE FOR INDUSTRY

The revolution in industry was comparable to that in agriculture. But there it made a more obviously purposeful impact – and not only in Russia, but also on the whole of the Western world. By a strange coincidence, the communists, through the First Five-Year Plan, seemed to be mastering their fate at precisely the same time as the rest of the world fell a hapless victim to the Great Depression. It was a grandiose, striking contrast – on the one hand, a nation in arms against poverty and insecurity; on the other, a world shaken hither and thither by economic collapse and catastrophe. This is an overstatement, of course, an overdrawn contrast. But it had sufficient validity to put Russia, a country standing only a decade earlier on the brink of dissolution, in the forefront of world economic development.

The heart of this effort was the Plan, a six-volume work, the result of two years' study by Gosplan. It went into operation on 1 October 1928, on a scale not easily analysed or described. The whole of Russia was hurled into a gigantic struggle to build socialism, to transform Russia from a backward agricultural into an advanced industrial country. Class A industries – coal, iron, steel, oil, and machine-building – were scheduled to triple their output; Class B industries, producing consumer goods, were to double their output. Overall, the gross output in 1932–3 was scheduled to rise to 236 per cent of the output of 1927–8. To support this effort the production of electrical power was to rise by 600 per cent. Almost no foreign capital, as distinct from foreign technicians, was available (owing to the reluctance of governments and banks to sanction loans to the U.S.S.R.). All was achieved through the ruthless accumu-

1. Frank Lorimer, *The Population of the Soviet Union*, Geneva (League of Nations), 1946, p. 121.

lation of capital, the ploughing back of surplus, and the limitation of personal consumption and amenities.

In this way were created the vast ironworks and blast furnaces of Magnitogorsk beyond the Urals, the hydro-electric plant on the Dnieper (under the direction of an American engineer, Hugh Cooper), tractor-works at Kharkov and Gorki, the Ukrainian industrial area based on the coal of the Donetz Basin, the iron-ore of Krivoi Rog and manganese and other mineral deposits, railways in Turkestan, oil refineries and pipelines in the Caucasus and Transcaucasia, machinery works in Smolensk and throughout the central Moscow industrial region. The Chelyabinsk tractor plant alone covered an area larger than all the old city of Chelyabinsk.

Side by side with this development went the vast enlargement of towns, not only of such old-established urban centres as Moscow, Leningrad, and Kharkov, but also of lesser centres such as Tashkent, Minsk, Vladivostok, Voronezh, Sverdlovsk, Novosibirsk. Living conditions were deplorable and standards of comfort and amenities non-existent, especially in the new towns. The only comparison is with the worst years of the Industrial Revolution in England. John Scott, an American engineer who worked for five years at Magnitogorsk in the thirties, writes:

I was going to be one of the many who cared not to own a second pair of shoes, but who built the blast furnaces which were their aim. I would wager that Russia's battle of ferrous metallurgy alone involved more casualties than the battle of the Marne. All during the thirties the Russian people were at war. . . . In Magnitogorsk I was precipitated into a battle. I was deployed on the iron and steel front. Tens of thousands of people were enduring the most intense hardships in order to build blast furnaces, and many of them did it willingly, with boundless enthusiasm, which infected me from the day of my arrival.[1]

Labour policy went into the melting-pot, as did everything else during these early thirties. The unemployment of the N.E.P. period suddenly gave way to labour shortages, which were in part overcome by sheer coercion, in part by a disguised

1. John Scott, *Beyond the Urals*, London, 1943, p. 9.

form of direction of labour, and in part by sheer patriotic upsurge. Clearly, the last was the most important impulse, at least in the early years of industrialization, and it was fanned and fostered by every device of mass communication. Not since the days of the First World War had such attention been paid by government propaganda to the common man in his role as producer. In any Russian newspaper of the time, it has been said,

the workers speak with their own voices and write with their own pens; on four pages of very poor paper, with very poor print, the vocal soldiers of industry shout themselves hoarse, with boasting, with exhortation, with criticism of failures, with challenges to socialist competition, with offers of 'tow-ropes' to less forward enterprises, with promises, with indignation.... Next we come upon a grave article upon the problems of technical construction in the coal industry.... Our correspondent complains of short production of coal in the Donetz Basin, there is absenteeism of labour on a large scale.... The repair of locomotives on the Murmansk line is unsatisfactory... A locomotive came back from the Volodga repair shop, after overhaul, with seventy defects.... Next we have a page devoted to agriculture, with a great headline across it: 'Quick collection of seeds shows Bolshevik leadership'....[1]

And so it went on – an increasing flow of criticism and exhortation, always and everywhere associating and identifying the worker with the national effort.

An essential aspect of labour policy in these years was education. There was no reserve of skilled labour to draw on. Men had to learn their jobs, or even be taught them while they worked. At the lowest level factories would themselves set aside facilities for teaching illiterate adults to read and write. At a higher level, facilities for technical instruction multiplied vastly. Technical institutes, colleges and universities, factory schools, all were pressed into service to provide new *cadres* of skilled workers.

On the other hand, precisely because the labour force was so undisciplined and untrained and had no traditional labour *mores*, the Soviet Government had to bring into play all sorts

1. Sir John Maynard, *The Russian Peasant and other Studies*, London, 1942, I, pp. 250–1.

of direct and indirect pressures. It enlarged the authority of managers and trade unions to discipline absentee or delinquent workers. Those who infringed employment regulations exposed themselves to criminal prosecution. The Government also made pensions, whether on grounds of retirement or medical disability, dependent on the worker's employment record. Similarly, inequality, differential rewards for skill, and extended piece-work created additional incentives in an economy lacking the profit motive.

At the end of it all, living standards had undoubtedly declined and the Plan had not been fulfilled. But Russia, just as certainly, ranked amongst the world's foremost industrial Powers. At a time when industrial production in the principal capitalist Powers had actually declined below the level of 1913, that of Soviet Russia showed an almost fourfold increase over the level of 1913.

CHAPTER 20

A Totalitarian Society

THE fear of war, the need to prepare for war, had been one of the mainsprings of Soviet industrialization. 'The industrialization of the country', said Voroshilov, the Commissar for Defence, 'predetermines the fighting capacity of the U.S.S.R.' This fear was not only an obsession born of past foreign intervention, it was not only the conviction of isolation and exposure, it also drew its strength from the Marxist analysis that capitalism and war were inseparable. Hitler's assumption of power could not but deepen this fear and give it manifest shape.

The fear of war, with its socio-industrial consequences, superimposed on the collective values inherent in Bolshevism, combined to create a society that was more and more totalitarian. The process began in earnest with the First Five-Year Plan and developed at an accelerating pace during the Second and Third Five-Year Plans. This is a familiar accompaniment of great national efforts.

Take first the adulation of the leader and the led. Ever since Lenin's body had been embalmed in a mausoleum in Red Square, before the Kremlin, to become an object of pious pilgrimage, a strong urge towards the consecration of charismatic leadership, beyond criticism and even beyond history, had grown up. With the presentation of Stalin as Lenin's successor, and all the more when Stalin exacted greater and greater sacrifices, the adulation of the leader sank to greater and greater depths. His full-page portrait first appeared in *Pravda* in 1929. Thenceforward, both visually and verbally, through every medium of modern publicity, the words and idealized countenance of Stalin were used to spur on and galvanize a population that was barely beginning to become literate, barely beginning to handle machinery, work at an assembly line, or take part in modern industry. In a similar manner, the drive for increased productivity led to the glorification of the producer, the coal-miner such as Alexey Stakhanov, say, who could hew ten tons where one ton was the norm.

Hand in hand with the State-inspired emphasis on the dignity of labour went the installation of a new respectability. There was a rehabilitation of virginity as an ideal, an end to easy divorce, the limitation of facilities for abortion, and the pro-clamation of the virtues of a stable family and married life. *Pravda* once wrote: '... A bad family man cannot be a good citizen and social worker.' In addition to exploiting the family as a unit of social service, the régime also sponsored the crea-tion of youth groups such as the Young Pioneers, the Kom-somol, and a number of part-time para-military organizations grouped together in the *Osoaviakhim*.

The new intelligentsia that developed in the thirties had none of the questing character of its predecessors. It accepted its values ready-made, perhaps with a certain amount of cynicism, an inevitable reaction after the hectic pace of the First Five-Year Plan. Its task was not to question the obligations and tasks handed down from above. These people were primarily technicians in outlook, concerned with achieving results and not necessarily with the type of result achieved. They grew up in an atmosphere steeped in Soviet patriotism, to which the uninhibited study of history was sacrificed, and were sedulously flattered as the bearers of a new civilization. They were the men who benefited most from the new opportunities and prospects opened up by industrialization. Those who were Bolshevik Party members allowed the old cut-and-thrust of debate in the Party's erstwhile animated sessions to degenerate into sedate affairs where delegates listened in silence to set speeches. They limited their intervention to 'hoorahs' and applause.

Altogether, the superfluity of politics and of public and political discussion was one prominent characteristic of these pre-war years. There simply was no need for them, since the goals and values of society were already in existence. The Soviet system, as it developed out of the turmoil of industrial-ization and Russia's general thrust into the twentieth century, gave no scope for even limited discussion of the nature or aims of the new society. The national effort was so comprehensive and inclusive that there could be no questioning of its aims or methods. What sort of revolution could there be? What sort of revolution had there been? These questions no longer mattered. They belonged to the past. What mattered now was the urgency of building socialism – even though the precise meaning of the word might be conceived in practical terms without ideological content.

The climax to the growing totalitarian nature of Soviet society came with the great purges of the middle and later thirties. Here, too, the unremitting hostility of the Germans and Stalin's failure to achieve any working *rapprochement* with the Western Powers made their contribution to a reign of terror unparalleled in modern Russian history. The closest analogy is with Ivan the Terrible's persecution of the boyars in the sixteenth century. Perhaps, in sober fact, Stalin *did* identify himself with Ivan at this time. (He complained to Cherkasov, the actor who played the title role in Eisenstein's film, that Ivan's fault lay in not annihilating *all* his enemies: 'This was a mistake,' said Stalin.) Then, too, the isolation of Muscovy and the threat and even reality of war had created a dark atmosphere of tension and suspicion, justifying the wildest crimes. But Stalin's purges did of course also reveal the existence of widespread dissatisfaction and oppositionist sentiment. To deal with this was the *raison d'être* of the purges. All the 'confessions' of conspiracy and treason were extorted by various forms of pressure and torture.

A revolver shot in Leningrad on 1 December 1934 set off the machinery of terror. Nikolayev, a young dissident communist, fired the shot; it killed Kirov, a close associate of Stalin, in charge of the Leningrad Party organization.

That same evening, according to Khrushchev's secret speech in 1956, Stalin caused the following directive to be issued:

I. Investigative agencies are directed to speed up the cases of those accused of the preparation or execution of acts of terror;

II. Judicial organs are directed not to hold up the execution of death sentences pertaining to crimes of this category in order to consider the possibility of pardon because the Praesidium of the Central Executive Committee U.S.S.R. does not consider as possible the receiving of petitions of this sort;

III. The organs of the Commissariat of Internal Affairs are directed to execute the death sentences against criminals of the above-mentioned category immediately after the passage of sentences.

This directive, Khrushchev added, 'became the basis for mass acts of abuse against socialist legality'.

Nikolayev and such of his sympathizers as could be detected were tried *in camera* and executed. Zhdanov replaced the dead Kirov in Leningrad and went ahead with purging the Party organization of politically dissident elements. But so far he and Stalin limited themselves to deporting the suspects.

This took place in the early part of 1935. Some eighteen months later matters took a sensational turn when no lesser personalities than Zinoviev and Kamenev, together with some others, were charged with no lesser crimes than collaborating with foreign Powers to overthrow the Soviet State. They worked, it was alleged, under the leadership of the exiled Trotsky. The accused men confessed and were duly executed. Now the reign of terror began in real earnest, spurred on by Stalin and Zhdanov. In a telegram to the Politburo they demanded that Yezhov replace Yagoda as head of the secret police. 'Yagoda has definitely proved himself incapable of unmasking the Trotskyite-Zinovievite bloc. The O.G.P.U. is four years behind in this matter. This is noted by all Party members. . . .'

The O.G.P.U. accordingly began to catch up on those four lost years. At the beginning of 1937 a further group of Old Bolsheviks trod the same path to death. They also confessed to incredible crimes of treason. Early the next year the purge reached out to the Red Army and swept away Marshal Tukhachevsky, Chief of the Red Army, and Admiral Orlov, Commander-in-Chief of the Red Navy; and in March 1938, a final group of twenty-one of the highest Soviet personalities, including Rykov, Bukharin, and Yagoda, was charged with collaborating with foreign Powers to dismember the U.S.S.R., overthrow socialism, and restore capitalism. They too were found guilty and executed.

The men in the dock were of course only the most notable victims. Unnumbered thousands of lesser people perished or were deported. Among the prominent victims were all the members of Lenin's Politburo (apart from Stalin himself and Trotsky), many ambassadors, most of the surviving Old Bolsheviks, the top leadership of the Red Army, the upper and middle levels of the Communist Party, and many members of the organization of Red Partisans. Those deported fell into the hands of a special department of the secret police set up in 1934. It bore the title: 'Chief Administration of Corrective Labour Camps and Labour Settlements.' By the end of the 1930s this body had organized a network of camps stretching across the north of European Russia and into the north-east of Siberia.

CHAPTER 21

Foreign Affairs

FROM ISOLATION TO THE LEAGUE OF NATIONS, 1921-34

DURING all these years of turmoil and upheaval on the home front, Russian foreign affairs pursued a *comparatively* untroubled path. Neither Chicherin nor Litvinov, the Commissars of Foreign Affairs, were ever members of the Politburo, so that the work of their department was not involved in the internal Party struggle to the same extent as was, for example, the pace of industrialization. Hence there is a continuity that was absent elsewhere. Foreign affairs, so to speak, were above controversy.[1]

But to speak of foreign affairs in the conventional sense is already to prejudge the issue. Trotsky, the first Commissar, thought his job would be nothing more than to issue a few revolutionary proclamations and then 'shut up shop'. Trotsky, like all the Soviet leaders, hoped and fervently expected that revolution in Russia would be swiftly followed by revolution in Europe – first and foremost in Germany. In 1918 the direct revolutionary appeal was comparatively muted. But in 1919, as the wars of intervention got under way, it was inevitably stepped up. 'In that year [i.e. 1919] we sent fewer notes to governments but more appeals to the toiling masses,' said Chicherin, Trotsky's successor at the Russian Foreign Office.

These appeals to the toiling masses went largely unanswered. As in Britain, for example, they were not much more than some slight brake on interventionist and anti-Soviet policies. They never went further than that. By 1920, not only had the Soviet republics in Hungary and Bavaria shown themselves to be ephemeral creations, but throughout Europe as a whole the revolutionary tide had clearly ebbed. There was certainly no reversion to the stability of the days before the First World

1. But see pp. 278-9.

War, but there was also no movement towards communism. The Bolshevik gamble on seizing power in Russia in the confident hope of world revolution had shown itself unjustified – at least for the present. What was more, certain very tentative beginnings had been made in the direction of normalizing relations with the Russian border states to the west – Poland, Finland, and the Baltic States. In this setting the Soviet leaders had no choice but to seek, however unwillingly, a further *rapprochement* with the capitalist Powers. At the end of 1920 this compulsion clearly emerged and the Soviets settled down to play the diplomatic game.

From now on, and until the rise of Nazi Germany, Russian foreign policy can be seen as the adjustment of revolution to the standards and attitudes of capitalist diplomacy.

Russian foreign policy, as it slowly developed, differed basically from Tsarist policy in that its aim was defensive. It had to defend the revolution. It had to keep the capitalist Powers at bay and prolong to the maximum the breathing-space that Soviet victory in the civil war had secured for the régime.

This was no easy task to fulfil. Understandably, the first two years of Bolshevik power had created a state of entrenched hostility on both sides of the fence. The major capitalist Powers, with the important exception of Germany, had financed and armed the forces of counter-revolution in an attempt to destroy the Soviet State and had thwarted the latter's attempts to secure a negotiated peace. The Bolsheviks, for their part, had spared no effort to instigate revolution, not only in the capitalist world but also in its colonial territories. There would seem, in 1920, say, no basis whereby a state of permanent irreconcilability might be avoided. Was the world irretrievably split?

That this did not in fact prove to be the case was ultimately due to two factors: the patent weakness of the Soviets, not only in themselves but also in their capacity to stimulate revolution elsewhere, and to capitalist disunity. (Only now, when these two factors no longer operate, does the conflict again appear irreconcilable). And it was from a combination of these two factors that a *modus vivendi* slowly developed.

The cardinal principle of Soviet foreign policy was to extend into peace the capitalist disunity that had saved the country in war. The aim must be to thwart the formation of any capitalist united front. 'Our foreign policy while we are alone', said Lenin, 'and while the capitalist world is strong consists ... in our exploiting contradiction.'

But what contradictions? The post-1918 world offered, it would seem, quite a choice. There was the conflict between France and Britain, the conflict between the United States and Japan, the conflict between France and the United States. But in actual fact only one conflict offered itself for serious exploitation: the contradiction between victors and vanquished in the First World War, between Germany and the Allied Powers.

But why should Lenin pick on Germany as Russia's foothold in the capitalist camp? He was in fact merely reading the signs of the times.

Our existence [he said] depends on there existing a radical divergence among the imperialist Powers on the one hand, and, on the other, that the victory of the Entente and the Versailles peace have made it impossible for the overwhelming majority of the German nation to live. The Versailles peace has created a position such that Germany cannot dream of a breathing-space, cannot dream of not being plundered ... naturally her only means of saving herself is by an alliance with Soviet Russia, whither they [sic] are directing their glances. They madly attack Soviet Russia, they hate the Bolsheviks, they shoot their communists like real genuine White Guards. The German bourgeois Government madly hates the Bolsheviks, but the interests of its international position impel it towards peace with Soviet Russia against its own wish.

Lenin hoped, of course, that he would be able to combine support for the German bourgeoisie in its struggle against the imperialists with a proof 'to those people, conscious of the bourgeois yoke, that there is no salvation for them outside the Soviet Republic'.

In other words, he had as yet by no means abandoned the aim of world-wide revolution. But this attempt to dance at two weddings at the same time soon showed itself a failure. The national aspect of a pro-German policy developed apace; the international aspect languished.

Lenin's diagnosis of the situation in Germany was remarkably accurate. It derived from the pro-Russian orientation of German policy ever since the collapse of November 1918. The Germans had refused to join in the blockade of 1919 imposed on Russia by the Western Powers; and in 1920, at the time of the Russian advance on Warsaw, they had refused to allow France and England to transport munitions across Germany in a hasty attempt to succour the Poles. It was then that Lenin noticed the emergence of a strange phenomenon in Germany – what he called 'a reactionary-revolutionary'. 'Everyone in Germany,' he noted, 'even the blackest reactionaries and monarchists, said that the Bolsheviks would save us, when they saw the Versailles peace splitting at all its seams, that it is the Red Army which has declared war on all capitalists.'[1]

Lenin's policy of *rapprochement* with the capitalist world had its first notable success in March 1921 with the signature of the Anglo-Russian commercial treaty. This, wrote Maisky, the future Russian ambassador to London, 'served as a door opening on to the arena of world politics'. Two months later this was followed by a Russo-German trade treaty. Its main feature was a German undertaking to recognize solely the Soviet Government as the Government of Russia. Later that year the Russo-German link was strengthened by the formation of a number of mixed companies, and the allocation of Russian mineral and trading concessions to a number of influential German industrialists. Before 1917, Leonid Krassin, the future Commissar of Foreign Trade, had combined underground Bolshevik activities with the St Petersburg representation of Siemens-Schückert. The excellent contacts that he enjoyed with German industrial circles were now of great use.

The climax came on the morning of Easter Sunday, 1922. This was the Russo-German treaty signed at Rapallo. It an-

1. Cf. also Churchill: 'The reactionary Germans would of course be delighted to see the downfall of Poland at the hands of the Bolsheviks, for they fully understand that a strong Poland standing between Germany and Russia is the one thing that will baulk their plans for [an imperialist] reconstruction and revenge.' *The World Crisis – The Aftermath*, London, 1929, p. 265.

nulled all mutual claims between the two countries; it re-established full diplomatic and consular relations, and it pledged the two Governments 'to cooperate in a spirit of mutual goodwill in meeting the economic needs of both countries'. The treaty greatly shocked France and Britain. Lloyd George feared the development of what he termed 'a fierce friendship' between the two 'pariah-nations' of Europe. The French even tried to secure the annulment of the treaty.

What did the treaty of Rapallo mean to Russian diplomacy? Its actual terms were innocuous enough. But there was certainly substance to justify the fears of the West. The Western Powers thenceforward lost the capacity to dominate events in Central and Eastern Europe. There could be no more question of squeezing Germany to the uttermost. There could also be no question of ever forming a united capitalist front against Russia. To Germany, Rapallo constituted a weapon in the struggle against Versailles. The bargain sealed in Rapallo obliged Russia to support the German nationalist struggle against Versailles. But it also obliged Germany to remain neutral in any concerted capitalist move. Come what may, there would always stretch before the western borders of Russia this vast buffer zone of neutralized territory.

On this basis there grew up a very close *rapprochement* between the two countries; closer relations developed, in fact, between these two states, the one capitalist, the other communist, than would ever again develop between the Soviet Union and a State with a differing social structure.

These relations were not merely diplomatic. A common Russo-German diplomacy marched hand in hand with a secret military understanding on rearmament. Negotiations in 1921–2 conducted under the authority of Trotsky, on the one hand, and the chief of the new German *Reichswehr*, General von Seeckt, on the other, led to the establishment in Russia of German factories for the production of poison gas, aeroplanes, and artillery shells. This arrangement enabled Germany to circumvent the disarmament clauses of the Treaty of Versailles and familiarized the Red Army with the most progressive Western military techniques. German credits for the later in-

304 THE MAKING OF MODERN RUSSIA

dustrialization of Russia also helped to consolidate the relationship.

Until 1933-4 the Rapallo treaty remained the sheet-anchor of Russian foreign policy. True, the smooth surface was ruffled from time to time: in 1923 by Russian support given to an attempted communist revolution in Germany; in 1924 by the German violation of the premises of the Russian Trade Delegation in Berlin, and later by the arrest of some German engineers in Russia. But not until 1931 were the Russians seriously compelled to review the situation.

In the meantime, there was considerable improvement in Russia's relations with the other Powers. In 1924 – paradoxically enough, hard on the heels of the attempt to foster revolution in Germany – the Soviet régime was suddenly recognized *de jure* by a whole host of important Powers. They included Great Britain, France, Italy, Norway, Sweden, Denmark, Austria, Hungary, and Greece. Only the United States, among the great Powers, still withheld recognition.

This relatively happy position suffered a severe shock in 1927. Overnight, it seemed, blow after blow suddenly rained on the Soviets. It began in March when the Italian Government, following pressure by Churchill and Sir Austen Chamberlain, ratified the Bessarabian Convention of 1920. (This sanctioned the Romanian seizure of Bessarabia from Russia in 1919.) The storm continued in April and May when Chiang Kai-Shek turned on the communists and their Russian advisers in the Kuomintang. This provoked an almost immediate breach in Russo-Chinese relations and virtually destroyed all Russian influence in China for more than twenty years. In May, a long period of Anglo-Russian friction culminated in the Arcos raid and the British rupture of diplomatic relations with Russia. In June, the Soviet ambassador to Poland was killed by an *émigré* monarchist. Later in the year, the laboriously erected Anglo-Russian Trade Union Committee was dissolved by the Trades Union Congress.

There resulted in Moscow a fear of war that was, to quote the French Ambassador, 'of an incredible intensity'. Only Germany stood apart from the Western Powers, and this enabled the storm to be ridden out. Then the worst could not happen and complete isolation be averted.

The natural sequel to this year of catastrophe was to intensify and accelerate every Soviet possibility of contact with the West and the non-Communist world generally. To the end of 1927 belongs Litvinov's celebrated policy of complete and immediate disarmament, a posture that would demonstrate the essentially peace-loving policy of the Soviets. Later Litvinov further developed his peace policy and brought about a whole network of non-aggression pacts as practical evidence of the Soviet will to peace. At this time the First Five-Year Plan was being introduced, and its concomitant upheavals, to say nothing of the enforced collectivization of the peasants, made the achievement of peace more desirable than ever before. There were non-aggression pacts with Latvia, Lithuania, Estonia, Turkey, Persia, Afghanistan, and later, even with Poland, and France. Russia also signed the Kellogg Pact of 1928 for the outlawing of war as an instrument of policy. By 1930 Soviet Russia was again, albeit with the significant exception of the United States, an accepted member of international society.

But there was a fly in the ointment – and that fly was Germany. For most of the previous decade the Soviets had given Germany invaluable backing in the struggle for the revision of Versailles. This lasted until 1930, when Litvinov, on the occasion of the final French withdrawal from the Rhineland, formally congratulated Curtius, the German Foreign Minister, 'on the restoration of the sovereignty of the German people'.

But this Russian policy had one danger that was very slowly making its influence felt. What if Germany should, with Russian aid, become so powerful as to turn against her erstwhile protector? Was Russia digging her own grave with the Rapallo policy? This point was raised in Moscow as early as 1929. An article in the authoritative journal of the Russian Foreign Office even spoke of a new German *Drang nach Osten* and pointed to the classical path of German economic expansion in South-East Europe and Turkey and Persia.

It was not, however, until the German attempt at a Customs Union with Austria in 1931 that Moscow saw the amber light.

This, as far as present evidence indicates, was a true turning-point. It demonstrated unmistakably that the Soviets were playing with fire in their support of German nationalism. But the Soviets did not at once jettison Rapallo. Instead, there ensued a twilight period of policy when Litvinov and Stalin cautiously and tentatively sought a *rapprochement* with Poland and France. The sequence of more and more right-wing chancellors in Germany – Brüning, Papen, Schleicher – gave added point and intensity to this new policy. The Germany that had once been the Soviets' foothold in the capitalist world was now slipping away. As a result, they sought to line up with France and Britain. It was again a continuation of the policy that Lenin had enunciated in 1920 – that of supporting one group of capitalist Powers against the other.

But until the very last minute the hope persisted that Rapallo might yet be saved. Until the beginning of 1934, Stalin and Litvinov were saying that 'we Marxists are the last people who can be reproached for allowing feelings to prevail over our policy.... Fascism is not the issue ... external policy counts, not internal'. But all this was an attempt to breathe life into a corpse.

FROM PEACE TO WAR, 1934–41

The evolution of Soviet foreign policy in the next few years contrasted strangely with the tone of home affairs. A period of terror, execution, and deportation at home was accompanied abroad by Litvinov's policy of collective security and the indivisibility of peace. At bottom lay the belief in the inevitability of war and the consequent urgency of staving it off by the exploitation of every diplomatic device. Stalin and Litvinov followed essentially the same policy that they had favoured before Hitler's assumption of power – that of preserving the *status quo*. But they did so in incomparably weaker circumstances. They had to deal simultaneously with a Western enemy – Germany – and a Far-Eastern one – Japan.

In the Far East the Soviets played a waiting game on the whole, buying off the Japanese, now with oil prospecting rights in northern Sakhalin and now with a fishing agreement. In 1935 the Japanese push into Outer Mongolia had to be bought off more dearly with the forced sale of the Chinese Eastern Railway at a small fraction of its real value. Even so, there was no peace in the Far East, and particularly not after the Japanese adhered to the Anti-Comintern Pact in November 1936. In 1937–8 Russo-Japanese clashes on the Manchurian border developed into a full-scale war.

Here there was little that diplomacy could do to help. It was different in the West, where Stalin and Litvinov could propound a policy that offered some hope of restraining the Nazis. But now the Soviets would reap something of the whirlwind that they themselves had sown in their support of German nationalism.

It was Hitler and not Stalin who broke off the Rapallo policy. He did so, after a year of wavering, in the most dramatic way possible: by concluding the German-Polish Non-Aggression Pact of January 1934. This brought the possibility of German aggression to the very borders of the Soviet Union. At the same time it put paid to all Litvinov's attempts at stabilizing the *status quo* in Eastern Europe. Although the Pact simply recorded the mutual aim of both Governments to effect 'a peaceful development of their relations', it was sufficient to align Poland with Germany. It thus dealt a severe blow at Soviet efforts to use amicable relations with Poland as a means to associating Russia with the Western Powers.

None the less, this effort inspired Soviet foreign policy over the next four or five years. In September 1934 Russia joined the League of Nations and was elected a permanent member of the Council. Henceforward, although there were tentative efforts to come to terms with Hitler, the Soviets were on the side of the angels. From this platform, day in, day out, Litvinov preached the necessity of a strong League, the principles of collective security, and the indivisibility of peace. In May 1935 the Soviets sought to give added emphasis to this policy by concluding pacts of mutual assistance, first with France and then with Czechoslovakia. In 1934, as part of the same policy, formal diplomatic relations had been opened with the United States for the first time since the revolution.

But the pacts with France and Czechoslovakia and the Soviet attempts to put teeth into the League had none of the success and developed none of the intimacy that had earlier characterized relations with the Weimar Republic. To a large extent the Soviets were imprisoned by their own revolutionary past. Stalin had long forsworn revolution, and in 1936 even issued a constitution embodying, ostensibly, all the democratic freedoms enjoyed by the Western democracies. But this could not allay the fears that he aroused. The Germans, through weakness, had not scrupled to sup with the devil, using the shortest of spoons. But even there the growing electoral strength of the German Communist Party in the early thirties had certainly helped to undermine the Rapallo policy.

The same process now took place in the West; and to it Stalin, unwittingly and paradoxically, made his own contribution. It was necessary to bring the Comintern into line with the new pro-League orientation, and this was achieved at the Congress of 1935. The Congress proclaimed the policy of the united front, calling on the international labour movement to rally to the defence of the Soviet Union and to the struggle against its enemies. This involved cooperation with any organization opposed to Fascism and the Fascist Powers. In France, for example, a by-product of the Franco-Soviet alliance was French Communist support for French rearmament. But the later accession to power of Popular Front Governments in France and Spain, and the general reappearance of communism as a political influence in other Western countries, further estranged these countries from association with the Soviet Union.

Russian foreign relations also suffered from the impact of the purges. What sort of State could it be in which such treachery was possible? Alternatively, what sort of a State could it be in which such false charges could be brought?

The overall result, therefore, was to leave the Soviet Union isolated. In isolation the Russians saw the progress of the British policy of appeasement and its fruit in German and Italian expansion – the reoccupation of the Rhineland, the conquest of Abyssinia, the *Anschluss* with Austria, the dismemberment of Czechoslovakia. The policy of collective security had not one success to chalk up. On each occasion that the Soviet Union proposed international action to deter aggression, or an international conference, as after the *Anschluss* and again after the final annexation of Czechoslovakia, it was shrugged off by the British.

Small wonder that Stalin felt justified in abandoning the quest for collective security, or any *rapprochement* with the West. In March 1939 he diagnosed the attitude of the Western Powers in terms of contempt. Their policy, he said, was tantamount to saying: 'Let every country defend itself against the aggressor as it will and can, our interest is not at stake, we shall bargain with the aggressors and with their victims.' But Stalin did not at once translate this discouraging diagnosis into practical politics. The switch, in the years immediately before and after Hitler's assumption of power, from Germany and Rapallo to the Western Powers and the League had been prepared most gradually, with every line of retreat covered and every line of advance carefully prospected. Similarly, it was only with the utmost caution that the decision was now taken to turn away from the West and seek an accommodation with Hitler.

The first very tentative move in this direction seems to go back to September 1938, at the climax of the Munich crisis. Stalin was scared by Munich, and the whole policy of appeasement, seeing in it an encouragement to Hitler to turn his armies against Russia. It was then that *Pravda* suddenly tarred the democracies and the Fascist states with the same brush, declaring that the Soviet Union saw 'no difference between German and English robbers'. But it was not until August 1939, almost a year later, that the irrevocable decision was taken to yield to German remonstrances, to cooperate in the fourth partition of Poland, and thus to unleash the Second World War. In the meantime, the reluctance, not to say refusal, of the British Government to enter into any alliance with Russia had become manifest. After the German occupation of Prague there had been the rejection of the Soviet proposal for a conference at Bucharest; then there had been the British guarantee to Poland without provision for further approach to the Soviet Government; finally, there had been the inability of the Anglo-French military mission to Moscow to tackle the question of how Russia might help Poland. But the Soviet Government, Stalin later told Churchill, was 'sure' that a mere diplomatic line-up of Britain, France, and Russia would not restrain Hitler. Such an agreement would need teeth.

As late as the beginning of August 1939, Schulenburg, the German ambassador to Moscow, was reporting to Berlin that it would take 'a considerable effort on our part to cause the Soviet Government to swing about'. But this effort was forthcoming. Perhaps it did not, by this time, have to be so considerable. On the night of 23–24 August, Molotov and Ribbentrop finally concluded the agreement that protected Hitler's rear for the imminent war against the West and 'bought in' the Soviet Government as his collaborators by sharing with them in the partition of Eastern Europe.

Russia had once again its foothold in the capitalist world. For the first year or so of the alliance with Germany, relations went smoothly enough. Collaboration did not only take an economic form, enabling Germany to defeat the British blockade; it also included the German use of certain Russian naval bases, a defeatist policy on the part of the Western communist parties, and even the delivery to Germany of several hundred European communists exiled in Russia. On this basis the Soviets were enabled to seize by force strategic outposts in Finland, to erect a defensive *glacis* along their western borders, and to annex the Baltic republics to the Soviet Union.

This latter strongly suggests that the Union had not overmuch confidence in its German ally – which was indeed the case. And this uneasiness was in fact considerably strengthened when the first of Stalin's miscalculations became manifest – the precipitate fall of France. He had seriously over-estimated French strength. It had been no part of his policy that Germany should so soon be freed from active campaigning in the West. On the contrary, he had hoped for, if not banked on, a much more protracted German involvement far from the Russian frontier. This would have given Russia a correspondingly protracted breathing-space to prepare for the inevitable struggle.

The consequence, therefore, was an outbreak of Russo-German friction immediately after the fall of France in the summer of 1940. But it did not reach its climax until November 1940. In a Berlin air-raid shelter, the issues were defined and the incompatible interests of the two allies made manifest. Their disputes turned largely on a classical source of Russo-German disharmony – the Balkans and South-Eastern Europe. This was precisely the area where the basis had been laid for the outbreak of the First World War. Was Bulgaria, for example, in the Russian sphere of influence or not? This was one of the crucial questions raised by Molotov, to the sound of exploding British bombs. Hitler, on the other hand, preferred to talk of the 'warm seas' of the Indian Ocean and slices of 'Greater Asia' in a general endeavour to divert Russia away from Europe and eastwards into conflict with the British. But Molotov was more concerned with what would happen *before* the British Empire was partitioned. There was not only Bulgaria; there was Finland, from which, Molotov proposed, Germany must withdraw her troops; there were the Dardanelles, where Russia desired to establish land and naval bases; there was 'the area south of Batum and Baku in the general direction of the Persian Gulf [which must be] recognized as the centre of gravity of the aspirations of the Soviet Union'.

All this far outstripped German willingness. Hard on the heels of these abortive negotiations followed Hitler's instructions for Operation Barbarossa: 'the German Armed Forces must be prepared to crush Soviet Russia in a quick campaign even before the conclusion of the war against England.' Until the last minute Stalin refused to give credence to the rumours and explicit warnings of a German attack. Diplomatically, he made every effort to patch up relations with the Germans. Even when the Germans turned against Russia on 22 June 1941 he at first ordered that their fire be not returned. Could it not be an error or a provocation by undisciplined German units? Would not the most pacifying answer be silence? This was the unedifying prelude to the Great Fatherland war.

CHAPTER 22

War and Peace

BETWEEN 4 a.m. and 5.30 a.m. on 22 June 1941, a new epoch in the history of the world began. German troops, joined by the whole of Fascist Europe, were moving eastwards for the kill. This is not the place to describe the immense struggle in any detail, merely to note certain of its consequences.

'War is merciless,' Lenin had said; 'it poses the problem with merciless clarity – either to perish or to surpass the leading countries.' By this merciless test, the Soviets emerged triumphant. In May 1945 Russia and Stalin enjoyed an apotheosis of glory and victory. Not only was the new Soviet empire becoming more and more a reality; also, and this was the truly noteworthy feature, the Soviet Union had survived what Stalin called 'the most cruel and the hardest of all wars ever experienced in the history of our Motherland. ... The point is,' he went on, 'that the Soviet social system has proved to be more capable of life and more stable than a non-Soviet system. ...' True, there had been a crisis of the régime in the winter of 1941–2; true, military weaknesses had been revealed; true, the loyalty of certain national groupings had been easily undermined; true, the bulk of the Russian forced labourers in Germany refused repatriation. Yet, for all that, the régime had withstood the severe arbitrament of war.

But at what a fearful cost! The Germans, taught by Fascist theory to regard the Slavs as sub-human, committed atrocity after atrocity against the civilian population, starved their Russian prisoners to death, and perpetrated manifold acts of wanton destruction. Russian casualties numbered perhaps ten million.

Material damage was equally extensive, above all in the areas west of the Volga. It has been calculated that one quarter of all Soviet property was destroyed – 17,000 towns, 70,000 villages and hamlets, 31,000 factories, 84,000 schools, 40,000 miles of railway track, not to speak of almost forty-five million horses, head of cattle, and pigs. When the war ended, about twenty-five million people in the western provinces had nothing to live in but wooden huts.

As against this, the German attack had brought Russian power into the heart of Europe – an unprecedented, incalculable phenomenon, and the source of interminable friction with the Western Powers. In Germany, Poland, and the Balkans, the Soviet Union was hard put to it to retain the gains of war in the immediate post-war period. In places, as in Iran, it even sought to extend them.

In these unpropitious circumstances of international friction and of economic devastation, the Russians had to turn to the back-breaking task of reconstruction. And once again this had to be accomplished largely from the country's own resources. Lease-Lend supplies ceased soon after the end of the war. Nothing came of the request for an American loan, desultorily discussed during the war. Worst of all, the Soviets received virtually no reparations from the Western Zones of Germany and all their proposals for four-power control over the Ruhr industries were rejected. Reparations from Eastern Germany, Manchuria, and supplies of manufactured goods and certain raw materials from the Balkans could by no means make good these handicaps. There was also a heavy price in political unpopularity to pay, especially in Germany.

In August 1945, the preparation of a Fourth Five-Year Plan was ordered. This was approved in March 1946. As before, the Plan put its emphasis on capital investment, particularly in electrification, irrigation works, and communications. Canals, power-works, hydro-electric plants, were its characteristic features. A special aspect was investment in the Volga and Ural areas and in Siberia. The Plan envisaged that production in the exposed western areas of the Union would be restored only to its pre-war level and not made the object of increased investment. In general terms, the aim was not only to reconstruct the economy but also to exceed pre-war levels of production and raise the material and cultural standards of the population. Housing did indeed take up a large part of available resources.

But Stalin was actually thinking much further ahead – a whole sequence of planned investments that would bring about a chimera of absolute security. In a famous speech in 1946, looking far beyond any immediately realizable targets, he was envisaging 'a situation where our industry can produce annually up to fifty million tons of pig-iron, up to sixty million tons of steel, up to 500 million tons of coal, and up to sixty million tons of oil'. Only in such conditions did Stalin consider that Russia could be 'guaranteed against all possible accidents'.

In the event, even to achieve the Fourth Five-Year Plan required an unprecedented degree of coercion, dragooning, and exhortation. But achieved it was. The output of heavy industry as a whole more than doubled between 1947 and 1950. The production of pig-iron, steel, electric power, and oil either doubled or came near to doing so. On the consumer front, food-rationing came to an end in 1947. Agriculture still remained backward, despite the most intensive efforts in the direction of reinforcing the collective farms. (Even today, with some forty-five per cent of its working population employed in agriculture, the Soviet Union is relatively inefficient. The comparative figure for the United States is twelve per cent.) Intensified scientific research during the Fourth Five-Year Plan also produced the first Soviet atomic bomb in 1949. By 1950, Russia had become the world's second industrial power.

As in pre-war years, this mighty industrial effort was accompanied by intellectual and cultural repression. These were the years of the Iron Curtain, when every effort was made to cut off communication between Russia and the outside world; when Soviet citizens were forbidden to marry foreigners; when there was a vast increase in the number of forced labourers; when great-Russian chauvinism was preached and practised; when xenophobia characterized almost every utterance of the régime. This outbreak of unreason far exceeded that of the earlier period of industrialization. It drew its strength both from the patriotic exaltation inevitably attendant on victory, and from the need to nerve the population to greater and greater self-sacrifice and endurance. Consonant with this policy, Zhdanov, Stalin's protégé in the Politburo, unremittingly excoriated in Soviet literary and cultural life any spirit of pessimism, introspection, disillusion, or dissatisfaction.

The supremacy of Soviet culture and the rejection of any fancied or alleged non-Russian influences were carried to absurd lengths. This campaign attacked without exception every group of intellectuals and spread its baneful impact to every soi-disant scholarly periodical. Its last phase came in 1949, when it was extended to 'rootless cosmopolitans' and went on to 'unmask' Jewish critics and men of letters by open denunciation and also by publishing their original Jewish names in parentheses (where these had been Russified or otherwise changed). The climax to the anti-Jewish aspect of the campaign came in January 1953, when a number of Jewish medical men were accused of having killed Zhdanov (who died in 1948), of planning 'to undermine the health of leading Soviet personnel', and of being agents of the Joint Distribution Committee, an American-Jewish philanthropic organization.

Happily, two months later, before a new purge could begin, Stalin was dead. His biographer thus sums up the latter phase of his career :

... immured ... in the Kremlin, refusing over the last twenty-five years of his life to have a look at a Soviet village; refusing to step down into a factory; refusing even to cast a glance at the Army of which he was the generalissimo; spending his life in a half-real and half-fictitious world of statistics and mendacious propaganda films ... seeing enemies creeping at him from every nook and cranny ... pulling the wires behind the great purge trials; personally checking and signing 383 black lists with the names of thousands of doomed Party members ... inserting in his own hand passages of praise to his own 'genius' – and to his own modesty! – into his official adulatory biography....[1]

Hardly was the dictator's body cold than his successors, headed first by Malenkov and then by Khrushchev, broke decisively with the irrationality of Stalin's régime. 'No panic and no disorder' – this cry went out from the Kremlin when Stalin died. There was none. Whatever may have gone on behind the scenes, public order remained undisturbed. In this setting, any

1. Isaac Deutscher, *Russia in Transition*, New York, 1957, pp. 37–8.

number of significant breaches with the past were made. The new leaders of the Politburo officially disavowed the 'doctors' plot'; they clipped the power of the secret police; they declared a wide measure of amnesty for the prisoners in labour camps; they abjured 'the cult of personality'. This was certainly no emergence from darkness into light; at best, it was into a twilight zone. None the less, it signified a break with the old régime. In one significant respect, however, Stalin's successors were able to build on the foundations of the planned economy that he had initiated and enforced – and that is in the creation of a Soviet Welfare State. Welfare had not been neglected, even during the worst pre-war years of rearmament, capital investment, and industrialization. To take medical services, for example: from 1928 to 1941 the number of general hospital beds rose from 218,000 to 661,900, maternity beds from 27,000 to 142,000, sanatoria and health resorts from 36,000 to 142,000. Similarly spectacular increases applied in the case of tuberculosis clinics, consultation centres for women and children, and venereal-disease centres.

But it has been left for Khrushchev, when presenting the Sixth Five-Year Plan to the Twentieth Party Congress in February 1956, to take up a theme that is clearly to be central in the Soviet economy of the decade. 'Now that we posses a powerful heavy industry developed in every respect, we are in a position to promote rapidly the production of both the means of production and of consumer goods. . . .' The speech is reminiscent in tone of Stalin's repeated declarations: 'Life has improved, comrades. Life has become more joyous.' But in actual fact, now that Soviet achievements in free education, in medical care, in the beginnings of a Soviet week-end, in improved pensions, wages, holidays, reduced working hours, and even housing, are at last, at long last being made actual, the words embody a measure of meaningful reality. It is no doubt a degeneration from the promise of the new world of 1917 to the promise of a long week-end in the 1960s. But there are worse things in life.

What effect has the first beginning of material abundance had on the life and movement of the intellect? The increasing centralization of power in the modern State and the increasing proportion of the national income at the disposal of governments has everywhere entailed a decline in freedom. The characteristic Western pattern of higher consumption standards for the masses combined with mounting expenditure on preparations for war exists also in Soviet Russia. And intellectual independence is being destroyed between these two phenomena. But if the first factor is as yet relatively undeveloped in Russia, the second is well ahead, if not materially, then morally so. Nowhere, therefore, has the resultant loss of freedom been as comprehensive. 'Strange fury to wish to assimilate ourselves to the rest of the world!' exclaimed Tchaadayev in 1836. 'What is there in common between us and Europe? The steam engine, that's all!' This remark is almost as true today as at the time it was made. Despite the undoubted intellectual ferment of recent years, so much has passed Russia by – the modern movements in philosophy, economics, sociology, psychoanalysis. If, as Lord Acton proclaimed, 'the pursuit of truth is the great object for which the life of man is prolonged on earth', then Russia, after a brief efflorescence in the nineteenth and early twentieth centuries, has once again become irrelevant.

The World and the Revolution

To talk of the irony of history is to use no empty phrase. If books have their fate, then so have revolutions. The revolution carried out by the Bolsheviks has taken a course no less different from the one they envisaged than the revolution itself differed from that envisaged by the Bolshevik's nineteenth-century precursors.

What, then, has happened? Is it possible to identify the objective meaning of the revolution, independently of the hopes and aspirations of its progenitors? Does it 'illustrate the bankruptcy of humanism', as Berdyaev argued? Or is it an attempt 'to transform the Old Russia into a new America', as Toynbee has said? At different times Russia has 'thrilled Europe with joy' as did the French Revolution: at the time of the First Five-Year Plan it embodied man's mastery of his fate; during the war it became Europe's hope – and after the war Europe's would-be executioner.

Does the chameleon-like phenomenon admit of closer definition? It is clear, first, that the revolution was much less revolutionary than it appeared to be. Before many years were out the revolution had not only conformed to the standards and customs of the surrounding bourgeois world, but had also resuscitated practices drawn from the Tsarist past. In some cases, such as literary controls, the revolutionaries went beyond their forebears. Later, the revolution would emulate the worst historical examples in its treatment of opponents and recalcitrant groups, its trials and its oppression.

But it is only *sub specie aeternitatis* that comparisons become meaningless. In the workaday world of history such points of comparison must be left to one side if the uniqueness of the revolution is to be assessed, all the more so because the revolution itself cannot be taken *en bloc* but only as a continuing and evolving process. How, then, can it be said, as Goethe said of the battle of Valmy in 1792, that from here and from this time on 'there opens a new epoch in world history'?

To this there seem to be two answers: a Russian and an international. In the perspective of Russian history the revolution undoubtedly brings to an end that spasmodic, recurrent theme of Russian history that began in the sixteenth century with the efforts of Ivan the Terrible to modernize and Westernize Russia. In this perspective, the revolution represents less a break with the past than an effort – and a successful effort – to take over and carry further the work of industrialization which the Tsarist régime had been unable to take beyond a certain point. Marxism came to Russia as a means for developing the country's resources in raw materials and manpower when Tsarism had failed to achieve this. 'Stalin was the architect, not of the modernization of a backward country, but of the completion of its modernization. Stalin was Witte's successor in a quite direct and technical sense.' [1]

It is hardly fortuitous, given the intimate connexion between war and industrial development, that Marxism, in the guise of industrialization, should come to Russia precisely at the time when the old régime proved its failure in war. The incapacity of the old régime to wage a war, let alone win it, relegated that régime to the dustbin of history, deprived it of its *raison d'être*. The new world of Marxism came into existence to overcome the inadequacy of its predecessor. To Russia, the slogan 'catch up and overtake the West' signified a rivalry in power. This particular contest seems now to approach its successful finale. Turgenev's evocation of 'holy Russia', sunk in a drunken stupor, 'her forehead at the Pole and her feet at the Caucasus', has gone for ever. Russia has at last thrust itself into a position where it is one of the two major world Powers. This achievement, for what it is worth, has been brought about in not much more than a quarter of a century. The Russia that was beaten by the Poles in 1920 had by 1945 become a global Power. There can be few parallels in history of latent strength being so quickly brought into play.

1. W. W. Rostow, *The Stages of Economic Growth*, Cambridge, 1960, p. 60.

The introduction of Western industrialization to Russia could not only build on the earlier efforts of the Tsars. It also drew strength from the ethos pervading much of Russian life – the notion of State service, the notion of collective wisdom, the notion of cooperative endeavour. Much of this is summed up in the Orthodox doctrine of *sobornost* – 'togetherness'. Individualist ideals, though by no means absent from Russian thought, were unable to compete with the manifold expressions of collective thinking and collective action. The individual seeker after truth or the self-oriented writer have generally been condemned by their Russian contemporaries.

Thus it was that Russian industrialization went hand in hand with strong collectivist tendencies. The process of large-scale industrialization has compelled any state subjected to it to develop a collectivist outlook, in the sense that it is only possible by virtue of the workers' self-identification with the necessity for the process. But it is the Soviets who have made this identification so intimate, not least through the opportunities offered to the non-Russian nationalities of the Union.

So much for the Russian significance of the revolution. What does it mean to the world at large? Most strikingly, perhaps, the fact that it has proved to be the first onset of a revolution that is still in progress. It is surely not fortuitous that the revolution of 1905 was followed within a few years by revolutions in Persia, Turkey, and China. A highly illuminating contribution to this theme was made by Bukharin in 1927. At this time Bukharin supported Stalin and the theory of socialism in one country as against the theory of permanent revolution; but to do this he had in fact to show that Stalin was still an internationalist and that he had not abjured international communism. Yet Bukharin's defence, for all its time-conditioned character and the adventitious circumstances of its pronouncement, has in fact shown itself to be a remarkable anticipation of the 1940s and 1950s.

The international revolution [he said in 1927] does not belong to the future, it is a contemporary fact. It is quite wrong, naïve, and stupid to conceive of the international revolution as a unique event, a 'world conflagration', which will simultaneously, as at the peal of a bell, at a command, break out everywhere. It is stupid to imagine that there exists a mystical predetermined 'hour' when 'His Majesty the Proletariat' will seize power. The international revolution is a gigantic process which will occupy decades.... The world revolution will complete its course only when it has everywhere won power.... The international revolution appears in different forms, takes on different shapes.... But we stand in the midst of international-revolutionary developments and only blockheads, the blind ... can look around them and ask: 'Where is this elusive international revolution, anyway?'

Today, more than three decades later, the basic truth of Bukharin's prophecy has been abundantly confirmed. This is particularly the case in Asia. In general, of course, the early Soviet leaders, denying their own Russian experience, looked for and sought to instigate revolution in the industrialized countries of the Western world. But there was also thought for Asia. As early as 1908 Lenin anticipated a conjunction between the proletariat of Europe and the peoples of Asia; and in later years the same theme was taken up at different times by Stalin, Bukharin, and Zinoviev.

The Russian revolution has shown itself to be in fact the first of a series of revolutions in the under-developed areas of the world, and one which has also helped to inspire its successors. It has been part of a world-wide revolution, one that is still in progress, recognizable for its unity, despite very significant local variations. (It is because the cold war, the normal struggle between competing Powers, has been superimposed on this world revolution that the problem of international diplomacy has become so intractable. To the normal diplomatic struggle, the revolution has added extended dimensions and instruments, such as foreign trade, attacks on the morale of the home front, etc.)

Russia, then, an under-developed area in the nineteenth and early twentieth centuries and very much a colony of the Western Powers, has been the first country to undergo the twentieth-century type of revolution. This revolution is primarily an essay in Westernization. Its aim, broadly speaking, is to enable a country to overtake in decades a lag of centuries.

The kind of economic development that took place in the West between the seventeenth and twentieth centuries is now being compressed into decades. That, at least, is the aim. Whether, and to what extent, the aim will be realized is another matter. Everywhere, from China to Peru, the slogan is planning, development, industrialization. The ferment of the revolution in Russia has set in motion a comparable ferment elsewhere – though it is not, of course, everywhere communist. It has shown that a strong state can so force the pace of capital accumulation and expansion as to create, within a relatively short period, a heavy-industrial base – a synonym for economic and consequently military power. Second, a government-controlled policy of investment can 'iron out' the trade cycle and do a great deal to ensure a fairly steady rate of expansion, avoiding the ups and downs hitherto characteristic of the Western economies. Of course, the stronger the state, the more able it is to accelerate the rate of capital investment and depress standards of consumption. There is, lastly, at the end of this process, the prospect of a rise in living standards. This is the carrot to justify the sacrifices. It is this prospect that the Russian revolution conjures up before the eyes of the masses in the underdeveloped countries of the world.

Further Reading

THIS short list of books is limited to works written in English or available in an English translation. Many of the titles listed themselves contain bibliographies which will serve as a guide to further study.

1. GENERAL

The Icon and the Axe by James Billington, London, 1966.

Lord and Peasant in Russia from the Ninth to the Nineteenth Century by Jerome Blum, Princeton, 1961.

Russia, A History and an Interpretation by M. T. Florinsky, 2 vols., New York, 1953.

Land of the Soviets by James S. Gregory, London, 1946.

A History of Russian Literature by D. S. Mirsky, London, 1949.

Russian Art by Tamara Talbot Rice, London, 1949.

An Outline of Russian Literature by Marc Stonim, Oxford, 1958.

A Survey of Russian History by B. H. Sumner, London, 1947.

Russia Absent and Present by W. Weidlé, English translation, London, 1952.

2. SPECIAL ASPECTS

(a) *Historiography*

Rewriting Russian History ed. C. E. Black, New York, 1956.

Modern Russian Historiography by A. G. Mazour, 2nd edition, New York, 1958.

(b) *Economic history*

Soviet Economic Development Since 1917 by Maurice Dobb, London, 1948.

History of the National Economy of Russia by P. I. Lyashchenko, English translation, New York, 1949.

Sergei Witte and the Industrialization of Russia by Theodore Von Laue, New York/London, 1963.

3. FROM THE BEGINNING TO THE END OF THE SEVENTEENTH CENTURY

The Beginning of Russian History by N. K. Chadwick, Cambridge, 1946.

The Urge to the Sea by R. J. Kerner, Berkeley, 1942.

Ancient Russia by G. Vernadsky, Yale, 1943.

Kievan Russia by G. Vernadsky, Yale, 1948.

Culture of Kievan Rus by B. D. Grekov, English translation, Moscow, 1947.

Mongols and Russians by G. Vernadsky, Yale, 1953.

Ivan the Great of Moscow by J. L. I. Fennell, London, 1961.

Ivan Grozny (i.e. Ivan the Terrible) by R. Vipper, English translation, Moscow, 1947.

Correspondence Between Prince A. M. Kurbsky and Tsar Ivan IV of Russia ed. and trans. by J. L. I. Fennell, Cambridge, 1955.

Russia at the Close of the Sixteenth Century by Dr Giles Fletcher and Sir Jerome Horsey, ed. E. A. Bond, Hakluyt Soc., London, 1856.

The First Romanovs by R. N. Bain, London, 1905.

Russia Under Two Tsars, 1682–1689: The Regency of Sophia Alekseevna by C. Bickford O'Brien, California, 1952.

4. EIGHTEENTH CENTURY

Peter the Great and the Emergence of Russia by B. H. Sumner, London, 1951.

Peter the Great by V. Klyuchevsky, English trans., London, 1958.

The Memoirs of Catherine the Great ed. D. Maroger, London, 1955.

Documents of Catherine the Great ed. W. F. Reddaway, London, 1931.

Catherine the Great and the Russian Nobility by Paul Dukes, Cambridge, 1967.

The First Russian Radical, Alexander Radischev by D. M. Lang, London, 1960.

Russia and Europe 1789–1825 by Andrei Lobanov-Rostovsky, North Carolina, 1947.

Palmyra of the North: The First Days of St Petersburg by Christopher Marsden, London, 1942.

On the Corruption of Morals in Russia ed. and trans. by Anthony Lentin, Cambridge, 1969.

5. NINETEENTH AND TWENTIETH CENTURIES
TO 1917

Napoleon's Invasion of Russia by Eugène Tarlé, English translation, Oxford, 1942.

The First Russian Revolution by A. G. Mazour, California, 1937.

Michael Speransky by Marc Raeff, The Hague, 1957.

The Third Section by Sidney Monas, Harvard, 1961.

My Past and Thoughts by Alexander Herzen, English translation, London, 1924.

From the Other Shore and *The Russian People and Socialism* by Alexander Herzen, English translation, London, 1956.

Road to Revolution by A. Yarmolinsky, London, 1957.

Studies in Rebellion by E. Lampert, London, 1957.

Russia and the West in the Teaching of the Slavophiles by N. V. Raisonovsky, Cambridge, Mass., 1952.

Portraits of Russian Personalities by Richard Hare, Oxford, 1959.

Alexander II and the Modernization of Russia by W. E. Mosse, London, 1958.

The Transformation of Russian Society ed. C. E. Black, Harvard, 1961.

Pioneers of Russian Social Thought by Richard Hare, Oxford, 1951.

Russia under the Tsars by S. Stepniak, 2 vols., English translation, London, 1885.

Red Prelude, A Life of A. I. Zhelyabov by David Footman, London, 1944.

Rural Russia under the Old Régime by G. T. Robinson, New York, 1949.

Mikhailovsky and Russian Populism by James H. Billington, Oxford, 1958.

Youth in Revolt by Shmarya Levin, New York, 1930.

Roots of Revolution by F. Venturi, English translation, London, 1960.

Tsardom and Imperialism in the Middle and Far East by B. H. Sumner, Oxford, 1940.

The Russian Marxists and the Origins of Bolshevism by L. H. Haimson, Harvard, 1955.

Memoirs of Count Witte, English translation, New York, 1921.

Continuity and Change in Russian and Soviet Thought ed. Ernest J. Simmons, Harvard, 1955.

Three Who Made a Revolution by Bertram D. Wolfe, New York, 1948.

First Blood – The Russian Revolution of 1905 by Sidney Harcave, London, 1965.

Russia in Revolution 1890–1918 by Lionel Kochan, London, 1966.

The Russian Empire 1801–1917 by Hugh Seton-Watson, Oxford, 1967.

Rationalism and Nationalism in Russian Nineteenth Century Political Thought by Leonard Schapiro, Yale, 1967.

Conservative Nationalism in Nineteenth Century Russia by Edward C. Thaden, Seattle, 1964.

6. SOVIET PERIOD

The Fall of the Russian Monarchy by Bernard Pares, London, 1939.

The End of the Russian Empire by M. T. Florinsky, Yale, 1931.

Russian Revolution 1917 by N. N. Sukhanov, ed. and trans. J. Carmichael, Oxford, 1955.

History of the Russian Revolution by L. Trotsky, English translation, London, 1934.

Formation of the Soviet Union by Richard Pipes, Harvard, 1954.

The Bolshevik Revolution by E. H. Carr, 6 vols. to date, London, 1950–9.

The Russian Peasant and Other Studies by Sir John Maynard, 2 vols., London, 1942.

Stalin, A Political Biography by I. Deutscher, Oxford, 1949.

Trotsky – The Prophet Armed by I. Deutscher, Oxford, 1954.

Trotsky – The Prophet Unarmed by I. Deutscher, Oxford, 1959.

Humanity Uprooted by Maurice Hindus, London, 1929.

The Socialized Agriculture of the U.S.S.R. by N. Jasny, Stanford, 1949.

In Search of Soviet Gold by J. D. Littlepage and Demaree Bess, New York, 1938.

The Communist Party of the Soviet Union by L. B. Schapiro, London, 1960.

Why Lenin? Why Stalin? by Theodore Von Laue, London, 1966.

7. FOREIGN AFFAIRS

Russia and the West under Lenin and Stalin by George Kennan, London, 1961.

The Soviets in World Affairs by Louis Fischer, 2 vols., Princeton, 1951.

Soviet Russia in World Politics by Robert Worth, New York, 1963.

Index